PERSPECTIVES ON BRITISH RURAL PLANNING POLICY, 1994-97

This is the first volume in a new series intended to provide a medium-term commentary on developments in British Rural Policy and Planning, based on the electoral span of a Government. The second volume, which is in preparation will thus examine the work of the Blair administration from 1997 to the next election.

It is accompanied by companion volumes for Europe and North America which however take a more systematic and sectoral approach.

Perspectives on British Rural Planning Policy, 1994-97

ANDREW W. GILG
Department of Geography, University of Exeter

Routledge
Taylor & Francis Group
LONDON AND NEW YORK

First published 1999 by Ashgate Publishing

Reissued 2018 by Routledge
2 Park Square, Milton Park, Abingdon, Oxon OX14 4RN
711 Third Avenue, New York, NY 10017, USA

Routledge is an imprint of the Taylor & Francis Group, an informa business

Publisher's Note
The publisher has gone to great lengths to ensure the quality of this reprint but points out that some imperfections in the original copies may be apparent.

A Library of Congress record exists under LC control number : 2002201917

ISBN 13: 978-1-138-33241-6 (hbk)
ISBN 13: 978-0-429-44666-5 (ebk)

Contents

Acknowledgements

I am grateful to Ashgate Publishing for allowing me to continue the series of books on Rural Planning Policy which I began in 1979 and which are more fully detailed in the Preface. In particular I would like to thank Sarah Markham who negotiated the transition from *Progress* to *Perspectives* and my co-conspirators in the *Perspectives* series: Keith Hoggart and Owen Furuseth. I would also like to thank Amanda Richardson, Anne Keirby and Valerie Rose of Ashgate for seeing this first volume through production.

I would like to thank the fairways of Axe Cliff Golf Club and Crans-sur-Sierre Golf Club, the pistes of Crans-Montana and the turf of Murrayfield for providing the relaxation without which no big task like writing a book can be accomplished.

Finally, this book is dedicated to the Scottish Rugby Union team who became five nations champions 10 days after this book was completed by defeating France 36 points to 22 in front of me and my wife, on her birthday, at the magnificent Stade de France in Paris, while on the next day Wales beat England by 32 points to 31 at Wembley.

Scotland team: Hilton, Bulloch, Burnell, Murray, Grimes, Pountney, Leslie, Reid, Armstrong, Townsend, Leslie, Tait, Logan, Murray, Metcalfe.

	P	W	D	L	F	A	Points
Scotland	4	3	0	1	120	79	6
England	4	3	0	1	103	78	6
Wales	4	2	0	2	109	126	4
Ireland	4	1	0	3	66	90	2
France	4	1	0	3	75	100	2

Andrew W. Gilg, Moorview Lodge, Exeter.
April 1999

Preface

This new series, with its companions in Europe and North America, is a successor to the *Countryside Planning Yearbook 1980-87*, the *International Yearbook of Rural Planning 1988 and 1989*, and *Progress in Rural Policy and Planning 1991-95*. These were annual volumes. However, the new series will take a longer more considered view, hence the change of title to *Perspectives*.

This first volume considers the period from the Autumn of 1994 to May 1997, when the Labour Party was elected for the first time since 1979. Autumn 1994 has been chosen as the start date, because this was the last period covered by the last edition of *Progress in Rural Policy and Planning* published by Wiley in 1995. Future volumes will be based on the electoral cycle, which should ensure a new volume every four to five years.

Individual chapters are structured around thematic sectors and take a descriptive rather than an analytical approach. However, reference will occasionally be made to the analytical devices employed in Gilg's *Countryside Planning* (Routledge, 1996). There are two core themes in each section. First, legislative change based on actual changes and proposed changes. Second, evaluation of policy effects by government agencies, pressure groups and books. The principal aim of the series is thus to update the reader on policy development and implementation and to make some general observations on policy impacts.

The book has been compiled from a wide variety of sources, notably: government policy documents; news stories from weekly publications like *Planning* and *Farmers Weekly* and from Press Notices and Quarterly Magazines issued by a wide variety of rural planning agencies; review copies of books from publishers; and reviews of books in learned journals. Learned journals however are not covered, since they are abstracted widely by databases like BIDS. In contrast, the evolution of rural policy at all levels and in all its complexity is not well covered, and so *Perspectives* hopes to fill a major gap in our knowledge of how rural policy is evolving.

1 Overview of Changes Affecting all Rural Planning Activities

The International Dimension

The Cork Declaration

In November 1996 the European Commission, in concert with the Irish Government, which held the EU Presidency, convened a conference which concluded with the issuing of the Cork Declaration: *Rural Europe-Future Perspectives*. The declaration had been triggered by three factors. First, the prospect of EU enlargement to the east. Second, the need to develop new policies in time for the next round of world trade talks due to start in 1999. Third, the obligation in the Maastricht Treaty to integrate environmental concerns into EU policies. According to observers (Lowe *et al*, 1996 and McDougal, 1996a) the declaration was a middle course between those countries like Germany and France who saw little need for further reform and those countries like the UK and the Scandinavian countries who wanted more trade liberalisation and the cost of the CAP much reduced. Although the declaration represented a move towards a wider set of objectives, one problem could be that DGXVI (Regional Policy) was not represented on the conference platform and this could indicate internal tensions over rural development strategy, notably the future of the structural funds.

The main principles of the declaration are set out below:

Rural preference Sustainable rural development must be at the top of the EU agenda and become the fundamental principle that underpins all rural policy.

Integrated approach Rural policy must encompass within the same legal and policy framework agriculture, economic diversification, management of natural resources, environmental enhancement and the promotion of culture, tourism and recreation.

Diversification Support for diversification of economic and social activity must focus on providing the framework for self-sustaining private and community-based initiatives.

Sustainability Policies should promote rural development which sustains the quality of Europe's rural landscapes, including natural resources, biodiversity and cultural identity.

Subsidiarity Rural policy must be as decentralised as possible, based on a bottom-up approach but within a partnership framework between all levels from local to European.

Simplification Agricultural policy and rural policy needs a radical simplification of legislation to ensure greater coherence, subsidiarity and flexibility.

Programming The implementation of rural policy should be through a single integrated programme for each region.

Finance Local financial resources must be encouraged to promote rural development.

Management More training for local government and community groups.

EU Legislation

The House of Commons Select Committee on European Legislation (1995) continued to provide extremely useful summaries and comments on the mass of documentation that pours out from the European Union. In the parliamentary year 1994 to 1945, 26 papers were issued which contained reports on the following topics. These can be found in HC 70 (94-95) with the exact paper number being given in brackets in the next few paragraphs which deals with the reports for 1994-95, 1995-96 and 1996-97.

The Committee accepted that the agricultural price proposals for 1995-96 (ix) would not be radical because the 1992 reforms were still working through, but they were nonetheless concerned at continuing high levels of expenditure, over complexity of the beef support system and lack of progress in reforming the unreformed sectors, such as sugar. The Committee also considered Arable Set-Aside (vi, vii) in particular, proposals for the 1988 five year scheme, and proposals to include land under agri-environmental measures and under afforestation schemes as eligible for the set-aside area were broadly welcomed. The Committee were however concerned (xxv) at the *ad hoc* and short term nature of decision making with regard to the yearly rate of set-aside and the way these had been based on not properly substantiated assumptions about world wheat prices. Proposals to prohibit the use of transplant seedlings grown conventionally in organic farming (xvii) were regretted by the Committee. Changes to the agri-monetary scheme in general and the proposal to freeze green rates were broadly welcomed in (i, iv, xiv and xxiii). Other agricultural issues to be considered were: the fruit and vegetable regime (ii); the EAGGF Guarantee section (ii); agricultural statistics (iv and xvi); beef intervention (iv); the sugar regime (v); and milk quotas (xix).

Other matters to be considered were: the LIFE programme (xxi) which is planned to grow by 12.5% to 450 million ECU in 1996-99 from 400 million ECU in 1992-95 in order to aid the creation of the Natura 2000 sites. The Committee doubted however whether this provided value for money. Proposed amendments to the 1979 Birds Directive (xvii) which related to the closed season for hunting migratory species were also considered. The Committee was concerned that proposed changes to the 1985 Environmental Asssessment Directive gave the environment disproportionate importance compared to economic development but accepted that the proposals were consistent with subsidiarity. Finally, the Committee also considered the environment and sustainable development (v); and the Structural Funds (xv).

In the 1995-96 parliamentary year the Committee (HC 51 (1995-96) 1996) examined the following topics. In agriculture the Committee were concerned that the price proposals (xiv) contained no proposals to reduce cereal or dairy product prices and no progress had been made on reform of the unreformed regimes, for example, fruit and vegetables. The Committee were also concerned at continuing high levels of expenditure and a reliance on quotas and maximum guaranteed areas which could fossilise production patterns. Turning to arable set-aside (i) (iii) and (xxiii) the Committee concluded that the system remained fundamentally flawed because constant changes in rates and penalties create instability and prevent long term planning by farmers. Reforms to the IACS system (xxii) were examined but considered to be too ambitious given that the existing scheme is far from perfect. The Committee did not have time to comment on the rushed proposals to modify the beef regime (xxix) by increasing some supports by 14% and in return cutting some cereal aid by 7%, in order to prop up the beef sector after the BSE crisis of Spring 1996. Proposed reforms to the fruit and vegetable regime (i) were however criticised for their reliance on Producer Organisations even though these were considered to be inefficient at delivering policy aims. Finally, the Committee noted the annual report on EAGGF expenditure (vi) and argued that the CAP could not be extended (vi) in its present form to the 10 countries of Eastern Europe as it was neither in their or western Europe's interest to do so.

In other areas, the Committee considered a draft directive on modifying the 1985 directive on Environmental Impact Assessment (iii) which they considered could conflict with UK Town and Country Planning legislation. The Committee also expressed concern that the list of new projects, for example, wind farms, that would or might need an Environmental Assessment were appropriate. A report on the Fifth Environmental Action Programme (xi) noted that there had been some but variable progress and so proposals (xv) were made to speed up implementation. Finally, the Committee considered draft ideas for a policy on the conservation of wetlands (i, vi and xiii) and argued that the definition of wetlands was too all embracing. Instead it queried whether a more targeted approach was needed and whether wise use of wetlands would better be achieved by bottom-up co-operation at the local level or top-down regulation via directives.

In the 1996-97 parliamentary year the Select Committee (HC 36 (96-97) dealt with a draft Council Regulation on organic production (i and xii). The Committee saw obvious merit in introducing a coherent structure, but thought that agreement might not be easy, even though many of the proposals reflected existing UKROFS (United Kingdom Register of Organic Food Standards) practice. Finally, the Committee expressed continued concern about the legality of proposals to amend the 1985 Environmental Assessment Directive (xvi).

The House of Lords Select Committee on the European Communities (1996a) produced a report on the enlargement of the European Union and CAP reform. The report was based on a strategy paper by the European Commission and broadly welcomed the analysis provided by the Commission. This argued that quotas and set-aside have no place in an enlarged CAP and that at the end of the expected transition phase farmers should be able to compete at world market prices. Compensatory payments should also be decoupled and be regarded as strictly time-limited adjustment aids. Reductions in price support should be coupled to policies for alternative forms of employment. Integrated Rural Policy is a welcome way of drawing together these considerations, but concepts such as cross-compliance are unhelpful. Finally, the report argued that agricultural payments for environmental services are particularly vulnerable to capture by interest groups and should not be seen as an alternative to price support but should only be made available under a clearly defined environmental policy.

Proposed reforms to the fruit and vegetable regime were also examined by the House of Lords Select Committee on the European Communities (1996b). The Committee did not like the proposals, in particular the proposed continuation of the intervention system and the enhancement of the role of producer organisations, because this would penalise the efficient producer, act against the interest of the consumer and continue to serve the interests of those who already produce the largest surpluses.

European environmental policies continue to grow in importance, as reported by Haigh (1995) and accordingly the House of Lords Select Committee on the European Communities (1995) examined the work of the newly created European Environment Agency and concluded that environmental legislation is not being fully implemented by member states and so proposed that the Agency could be given the powers of an inspectorate to visit member countries, and verify the accuracy and consistency of national methods of data collection, collation and analysis.

Environmental information was also the subject of a House of Lords Select Committee on the European Communities (1997) report on the implementation of the 1990 Directive on Freedom of Access to Information on the Environment which came into force in the UK in 1992 under the Environmental Information Regulations. The Committee concluded that the Directive has had some success in improving access to information and in stimulating public involvement in environmental protection. However, it is not achieving its full intended impact and so the Committee recommended that the EU Directive should contain a more forceful assertion of citizen's rights and a strong presumption in favour of openness.

In the UK the report noted that government departments and public bodies had made considerable efforts to promote openness, although some bodies still tried to evade the regulations by a number of ruses, for example by labelling documents as draft (Milne, 1997a). The Committee thus recommended that the Department of the Environment should draw up a list of bodies (not necessarily exhaustive) covered by the UK Regulations and should revise its guidance about exemptions relating to confidentiality, deliberations, internal communications and incomplete information, in order to guard against further misuse of possible reasons for non-disclosure of information. For those wishing to find environmental information the Department of the Environment issued *A guide to using public registers of environmental information* as a free document to the public in 1995.

Another European perspective is provided by Hoggart, Buller and Black (1995) in *Rural Europe* which asserts that the environment rather than agricultural use or low density is becoming the central defining feature of rural areas. Nonetheless, in spite of globalisation and the CAP, rural diversity is still more common than rural uniformity across Europe.

The Three Rural White Papers

The White Papers were announced in October 1994 and were to be the first countryside policy statement since the famous wartime Scott report of 1942. The decision to publish three papers, one for England, Scotland and Wales was heralded to be significant, as was the decision for the English paper to be produced jointly by the DOE and MAFF with close co-operation from other departments. Views were invited from a wide cross-section of organisations in November

1994 (see for example some publications later in this chapter) and these were analysed by Traill-Thomson (1995) from which it was concluded that the market ethos of the Government was at odds with the development of a rural policy. This conflict meant that it was virtually inevitable that a worthy but largely inadequate document would be produced, along the lines of '*Our Common Inheritance*', the 1990 statement on the environment and its annual successors.

The English White Paper

This was published in November 1995 (Cm 3016). It provided a sound, albeit bland, account of the issues facing rural England in 4 major sections which dealt with: Government and people; working in the countryside; living in the countryside; and a green and pleasant land. It stated six principles for the future of the countryside:

- the pursuit of sustainable development;
- shared responsibility for the countryside as a national asset, which serves people who live and work there as well as visitors;
- dialogue to help reconcile competing priorities;
- distinctiveness, approaching rural policies in a way which is flexible and responds to the character of the countryside;
- economic and social diversity; and
- sound information as the basis for effective policies.

It concluded with a checklist of action points. These were either exhortations, e.g. using cars less, or restatements of existing policies. The only new measures involved a new Rural Business Class to encourage rural enterprise and rate relief for village shops, and promises to introduce already announced changes, e.g. protection for important hedgerows. The immediate reaction was a guarded welcome for the general tone of the White Paper but disappointment about the lack of specific measures. Conservation groups and rural businesses were on the whole more pleased than the farming community.

An eclectic reaction (Potter, 1996) was provided by the House of Commons Environment Committee (1996a) which: argued that the relationship between PPG7 (countryside and the rural economy) and PPG13 (transport) should be reassessed; doubted if the target of doubling the woodland cover in 50 years was achievable; criticised the lack of targets indicating practical ways to carry forward the White Paper process; and expressed concern that the recently published countryside character map was no substitute for statutory protection.

The Government issued a report on progress made one year after the White Paper in 1996 (Cm 3444) which noted that legislation had been introduced to provide rate relief for village shops, and that progress had been made on revising PPG7, drawing up draft legislation for protecting hedgerows and had relaunched an expanded Countryside Stewardship scheme.

More considered reactions were contained in a number of review articles. For example, Hodge (1996) concluded that the emphasis on small, individual, albeit attractive initiatives at the expense of a more thorough analysis of the underlying forces which in practice determine conditions in rural England, and the lack of any attempt to draw out the inter-relationships between rural issues meant that an opportunity had been missed. Murdoch (1997) argued that governments have retreated from intervention and now stress how little government can do, so that the White Paper emphasised the role of self-help and self-reliant communities who can seek partnership from the state if they so desire. These two themes are reiterated by Potter (1995) who reported that the White Paper repeatedly extols voluntarism and community action and is an incrementalist's charter full of eye-catching initiatives heavily steeped in the language of the enterprise culture, but not as the Government claimed 'a new vision' about how to turn the countryside into a living, working place without harming the environment. The intellectual vacuum surrounding the White Paper was identified by Sidaway (1996) in contrast to a report by the National Trust (1995) which contained a refreshing debate about all aspects of the countryside. In contrast, Elson (1996) welcomed the White Paper as providing a useful starting focus for an annual process of auditing and co-ordinating rural policy.

The Scottish White Paper

The Scottish White Paper (Cm 3041, 1995) was a very different document to the English one in that it contained one major new policy, based on building three partnerships: Scottish Local Rural Partnerships; the Scottish National Rural Partnership; and the Scottish Rural Partnership Fund. These were welcomed by Moir and Watt (1996) especially their derivation from the integrated ethos pioneered by the *Rural Framework* studies (Lloyd, 1996) carried out before the White Paper which used a non-sectoral and integrated approach to rural issues. Other policies for rural Scotland were set out under the core themes of espousing the voluntary principle where possible and achieving sustainability as the ultimate aim. In common with the English White Paper the text was stronger on style than substance, painted too rosy a picture, and failed to provide a vision for a future Scottish rural landscape (Bryden and Mather, 1996).

The Welsh White Paper

The Welsh White Paper (Cm 3180, 1996) was more similar to the English paper in that it contained a number of similar measures, some of which reflected the special conditions of Wales, for example, proposals to relax planning controls to allow housing for relatives of retired farmers within the curtilage of the farmstead. However, the Paper was strongly criticised by the Labour Party who launched an alternative White Paper which proposed merging the Welsh Development Agency, the Development Board for Rural Wales and the Land Authority for Wales (Davies, 1996). Green (1996) described the document as the non-event of the year because it contained no new commitments, contradicted itself across chapters and emphasised the Welsh Office's *laissez faire* attitude to the environment.

Public Policy in General and Rural Planning in Particular

General Government Policy

The value for money, performance target ethos inaugurated in the 1980s has been continued into the 1980s as exemplified by Cm 3179 (1996) which also added new concerns about 'sleaze' and thus a need to vet propriety as well. In spite of the general antithesis to public bodies Cm 3557 (1997) concluded that the overall performance of public bodies was good. In parallel with these assessments, three annual audits of 'Next Steps Agencies' Cms 2750, 3164 and 3579 (1994, 1996, and 1997) were issued. These provided audits for ADAS (the rump research arm of the former advisory body), the Intervention Board for Agricultural Produce, and Forest Enterprise and noted that the Farm and Rural Conservation Agency (an offshoot from the former ADAS) and the Forest Service were under consideration for 'Next Steps' status.

Political Parties, the Professions and Pressure Groups

In 1995 the Labour Party issued *A Working Countryside* which according to Bullen (1995a, and *Ecos* 16 (2) 1995, pp. 77-8) promised to reform the CAP, expand organic and environmentally sensitive low input farming and improve the rural economy. In more detail all intervention buying and export subsidies would be abolished, there would be a 'right to roam' consistent with agricultural practice, services and employment would be boosted by a business development bank, planning permission in otherwise 'no-go' areas could be given for houses for renting by local people, no more sales of Forestry Commission forests, increasing tree cover by 50% by 2010, greater protection for protected wildlife habitats, and effective legislation to protect hedgerows would be introduced.

A year later the Liberal Democrats (*Farmers Weekly*, 2 August 1996, p. 13) issued *Country Lives and Country Landscapes* which among other things called for MAFF to lose responsibility for food and the establishment of an independent Food Commission to take over this role.

During the 1997 election the manifestos of the three big parties made the following promises (McDougal, 1997a). All parties wanted the CAP reformed with the Liberal Democrats advocating new direct management contracts but the other two main parties favoured modifying existing support systems. The Conservatives wished to set up a food safety council within MAFF, but the Labour Party and the Liberal Democrats wanted an independent food standards agency and a food commission respectively. Labour promised a cabinet minister for environmental protection, while the Conservatives wished to continue polluter-pays policies. The Conservatives promised additional funding for Countryside Stewardship with more cross-compliance, the Labour Party favoured organic farming and green premiums, while the Liberal Democrats offered direct management contracts for farmers to farm in environmentally friendly ways. In Forestry the Labour Party promised to stop sales of Forestry Commission land, while the Conservatives wished to double the forest area over the next 50 years. Finally, Labour offered a 'Right to Roam', the Conservatives wanted to encourage access by agreements and the Liberal Democrats supported grant aid for farmers to meet the environmental costs of access. In a commentary on the three manifestos Levett (1997) criticised the Conservatives for not facing up to their poor environmental record, Labour for failing to grasp the nettle of increased public expenditure on the environment, but praised the Liberal Democrats for a well crafted set of policies that would make a difference in the highly unlikely event of them being elected.

The RTPI, RICS, CPOS and the DPOS (1995) in a report *Tomorrow's countryside: A rural strategy*, which was intended to influence the rural White Paper made the following main points.

Agricultural policy

- the CAP should be extended into an integrated rural policy;
- rural strategy plans should be drawn up by local authorities, in consultation with the community, business and interest groups;
- agricultural product support should be cut, but kept at a level which leaves prices close to an open market price;
- funds should be shifted from agricultural production to environment measures;

- planning policies should allow for non-agricultural activities consistent with environmental aims to generate rural employment.

Sustaining rural communities

- better rural transport is needed;
- an adequate supply of affordable rural housing, and land allocation for that housing, is essential;
- village shops and post offices, schools and community centres should be sustained by a variety of means.

Conserving the environment

- measures are needed to promote and fund agri-environmental schemes;
- demand for primary minerals should be reduced by encouraging recycling;
- use of chemicals must be minimised.

Delivering the strategy

- a Cabinet committee for rural affairs should be set up;
- co-ordinated regional guidance should be readily available;
- rural strategy plans should be promoted and implemented;
- local guidance at parish/community level should be accessible.

Some commentators welcomed the report as a pragmatic and positive approach while others said it was 'Mom and apple pie' (Morrison, 1995) or a bland poorly co-ordinated wish list (*Ecos*, 16(2), 1995, p. 78).

Another document issued in an attempt to influence the rural White Paper was *Towards a Rural Policy: a vision for the 21st Century* published by the Country Landowners Association (1995a), which contained a 25 point action programme stressing the importance of deregulation. The document also called for a government department for rural affairs which would take control of forestry, wildlife, rural planning, flood defence, water supply, landscape conservation, environmental protection, pollution control and extraction of minerals (Young, 1995). The document feared that landscape designations could be misused by the planning system to constrain rural enterprise and thus called for more environmental 'stewardship' incentives to protect wildlife habitats for which farmers would have to reduce their reliance on subsidies. Turning to access the report argued that access was better ensured by voluntary agreements than by any statutory 'right to roam' and that country sports must be recognised for their role in the rural economy and environmental management. Finally, the report called on planners to be less restrictive over rural housing in order to meet the needs of the estimated 37,000 households in need and the 16,000 rural homeless and for schemes to promote jobs, rural transport and services in order to encourage the rural economy (Bullen, 1995b).

The CLA (1995b) also published a *European Rural Policy* which called for a phasing out of set-aside and production controls and shifting farm support towards environmental and socio-economic objectives, because production controls encourage inefficient farmers, hold back efficient farmers and increase overall costs to the industry. Farm support should therefore be frozen at current levels and decoupled payments introduced to meet environmental objectives, based on land stewardship and a less regulatory environmental policy. In addition, according to Cheesley (1995) the report also stressed the need to maintain regional social aid so that small and medium-sized enterprises could be encouraged to create new jobs in rural areas, using EU structural funds.

Rural Planning Agencies

The Countryside Commission (1996a) in a post White Paper document, *A living countryside: a summary of our strategy for the next ten years*, set out seven main themes:

*We will encourage local pride...*so that local communities and public organisations strengthen the special character of each part of the English countryside. Our work on this theme will include:

- the 'Countryside Character Programme', which looks at the nation's landscapes; and

*We will promote sustainable leisure activities in the countryside...*to meet changing leisure needs in a way which respects places, and benefits local people. We will do this by developing new policies, and work which will include:

- creating 'Greenways'-safe and convenient local links between the town and the countryside;
- completing our campaign to have all rights of way in good order by the year 2000; and
- further work in the field of 'green tourism'.

*We will achieve long-term benefits from farms and woodlands...*by finding a sound basis for stewardship of land. We will do this by continued policy development, and by:

- working to double the area of woodland in England; and
- promoting countryside products from land managed in ways which benefit landscape, wildlife and recreation.

*We will plan for sustainable development in the countryside...*to show how the principles of sustainable development can be applied in practice. We will do this by continued policy development and other work, including:

- experiments and advice on managing transport in the countryside.

*We will provide better information about the countryside...*to help our partners and the public as they make decisions. We will do this by:

- looking for better ways of supplying the public with information on the countryside;
- extending our work as an expert source of information about the countryside to other organisations.

*We will protect and promote the areas of finest landscape...*to keep the essential qualities of the National Parks, Areas of Outstanding Natural Beauty and Heritage Coasts. We will encourage the work of the people who manage them.

*We will improve the countryside around towns...*to create new landscapes and new opportunities for recreation in parts of the countryside closest to where people live, by:

- pressing forward with the programme of 'Community Forests'.

The document followed on from an earlier consultation document (Countryside Commission, 1995a) which had set out five challenges: how to conserve and enhance the quality and diversity of the countryside; how to secure balanced, multi-purpose farming and forestry; how to promote sustainable development in the planning and management of the countryside; how to improve and extend opportunities for people to enjoy the countryside; and how to encourage understanding of the life and work of the countryside.

Reviews of General Rural Issues

Murdoch and Marsden's (1994) *Reconstituting Rurality* is the second book to arise from the ESRC funded Countryside Change Project. The first book which was reviewed in Volume Three of *Progress in Rural Policy and Planning* (pp.121-2) set out revised perspectives for reviewing rural areas but contained no new empirical material. This book contains detailed research findings in relation to the land development process using the concept of middle class domination of decision making as the main explanatory concept and the case study as the main methodological tool. The county of Buckinghamshire is used almost exclusively for the research. The book is divided into 10 chapters. The first and the final chapters discuss the context in which land development takes place, while the intervening eight chapters provide empirical research findings.

In more detail Chapter One develops the idea that classes form themselves at particular places and at particular times, and that in rural areas they have used the planning system to mould a new vision of rural places based around a vision of the 'traditional village community'. This theme is developed in Chapter Two which analyses a questionnaire survey of households to reveal a strong anti-development stance among village residents. Chapters Three and Four develop this theme by demonstrating how the plan making and development control systems have been used to construct exclusivity and how villages are increasingly being populated by middle class residents. In the wider countryside, although agricultural land is becoming increasingly surplus, farmers are unable to find alternative uses, and Chapters Five, Six, Eight and Nine demonstrate how a climate of opinion originally in favour of diversification out of agriculture has been increasingly diverted to a culture which is only in favour of developing traditional buildings and land uses. These chapters use a number of case studies which focus on golf courses, reuse of farm buildings, and waste disposal to make their point. The special case of mineral exploitation is highlighted in Chapter Seven which is perhaps the most interesting of the chapters.

The final chapter uses Massey's concept of class development taking place in those localities where the actors involved play out their roles. In this case it is concluded that middle class in-migrants have sought to find, or if it could not be found, have created, traditional village lifestyles in rural Buckinghamshire. To do this they have used 'acceptable' concepts of peace and quiet and landscape preservation, rather than 'unacceptable' notions of excluding less well off people or people who do not conform to being white, English, family-orientated, and middle class. These findings are not of course new and the authors acknowledge that this is a long term process that has been studied by Pahl in the 1960s, Peter Hall and his co-workers in the 1970s, and John Short and his co-workers in the 1980s.

In developing the thread into the 1990s the authors are to be thanked and also to be congratulated on producing a very easy to read and jargon-free account based on a free flowing narrative style. This book will thus be really useful to students studying how the planning system actually operates, and to those students who increasingly wish to research dissertations using the concept of planning as a struggle between key actors. This book provides an ideal model, notably in the Mineral Planning chapter, which, as with most of the chapters, uses a three fold division. First, setting the context within which the struggle takes place; second a series of case studies showing how actors use the context, but also realise that the context itself might need changing; and thus third, the extent to which individual struggles have led to the context being changed.

In their words the process is as follows: '*These actors may be seeking to reproduce particular organisational patterns (part of what we see as the "context") but they may also be seeking to challenge such patterns. Sufficient challengers, commanding sufficient resources, can change the context irrevocably, setting the stage for the next "round" of contestation.*' (p.185)

In conclusion, this book by implicitly emphasising the incremental nature of planning policy development, albeit set within the context of a dominant middle class has reinforced rather than revised our view of rural planning policy. The authors are to be congratulated on producing such a useful description of rural planning as it is actually practised.

Michael Winter's (1996) *Rural Politics* is a very welcome introduction to the literature in that it provides a long term and interdisciplinary perspective on the development of rural policies from the viewpoint of those involved in making and influencing policy. Thus it provides not only a documentary on policy change but an explanation of why these policy changes occurred in the way they did.

The book is written from an unashamedly historical perspective dating often from the 19th century, although most of the text concentrates on developments since 1945. Three themes dominate the text: first, the incrementalism of policy change; second, the dominance of corporatism in policy making and delivery; and thus third, the importance of understanding the policy networks of groups and individuals which structure the policy process. The book is divided into three parts. The first part examines the various possible approaches to policy change. The second part discusses agricultural policy, while the third part examines rural environmental policy.

The first part is a very useful summary of the policy process and how the subject of policy analysis has developed the concept of policy networks within which key decisions are made. This concept is more empirical than some of the more rigorous abstract models, but reflects the real world of policy making which is messy, incremental and often influenced by chance factors. Using this approach, Winter describes how policy making works in Britain, and then discusses the problems of not only implementing policy but also how to assess the impact of policies which until recently had been very neglected.

The second part focuses on agriculture and describes the rise and partial fall of agricultural corporatism and the close links between MAFF and the NFU between 1945 and the 1990s. The analysis however also includes a useful section on policy development before 1945 which is often neglected. The analysis is also useful because it includes the views of those people engaged in the process, and thus provides greater explanatory depth than those accounts based on a more logical interpretation of events from economic and statistical trends. The third part widens the focus out into rural environmental policy.

The first chapter traces the origins of environmental concern and again extends the discussion back to the 19th century. The second chapter looks at protecting landscapes, habitats and wildlife and in a reflection of the field jumps from one piece of policy to another. One constant theme however is the assertion of property rights by the NFU and CLA via the mechanisms of compensation payments and the voluntary principle and self regulation. This is followed by a chapter on the greening of agricultural policy which includes a perhaps overlong section on advice and FWAGs which reflects the recent research interests of the author. The fourth chapter examines agricultural pollution and again highlights the piecemeal and fragmented nature of policy making and delivery. The fifth chapter examines forestry and highlights the closed nature of the policy community which had made forestry farming's rich relation, although it is now considered to be beleaguered. The final chapter briefly examines five themes: restructuring; sustainability; reforming the CAP; Conservative policies; and the farm lobby.

The book is thus an important new contribution to our understanding of how policy evolves, but it is however only a partial contribution to our understanding of wider rural politics. In particular, there is virtually nothing on the Town and Country Planning system or recreational policy. There are also major omissions. First, there is no reference to various previous attempts to chart rural policy, notably Andrew Gilg's 1978 text on *Countryside Planning* and his series of *Yearbook* and *Progress* volumes which contain very similar information to large sections of Winter's book. Second, there is little attempt to analyse the policy impact of policy from the viewpoint of published research, except when policy changes are being explained. Third, some rural organisations are almost entirely ignored, notably the Countryside Commission which has no references at all. Fourth, the book makes little use of the models set out in the early chapters in analysing policy change.

This book is thus a very useful review of policy change in **agriculture** but not a review of wider rural politics. However, apart from the rather misleading title it provides an extremely useful complementary text to Andrew Gilg's totally new text *Countryside Planning* also published in 1996. The two books should be read together in order to provide a very clear explanation of how rural policies have evolved and what effect they have had.

In more detail Gilg's (1996) *Countryside Planning* is a totally new edition of the first edition which was published in 1978. Since then there have been three major changes. First, the thrust of policy has been markedly altered by 16 years of Conservative Government between 1979 and 1995. Second, agriculture has lost its primacy in rural planning as a result of European food surpluses, and this has presented other land users with an opportunity to fill the partial void. Third, there has been an explosion of literature on rural planning topics, notably from academia and the pressure groups.

There was thus a need to totally reconceive the concept used in 1978. The one central concept that has survived however, is that of incremental and evolutionary change as the main explanatory factor. The two editions can thus be usefully read in parallel, since, the first edition not only provides more detail on the 1945-75 period, but also provides a contrast with the themes and topics that were important then, but have since fallen by the wayside. In particular, the first edition included much material on planning techniques, but the literature has largely fallen silent on this topic since then, in favour of debates over research methodologies. Indeed, while empirical analyses dominated the first edition, this edition divides research material by the methodological and theoretical approach adopted by the researchers. The analysis is thus more discursive than before.

Three major forms of analysis are used in the text, namely:

i) Traditional meta-narrative empirical analyses based on the assumption that there is a policy effect, and that the impact of policies can be measured quantitatively in a cause and effect relationship, with for example winners and losers being identified usually by statistical or econometric methods;

ii) Structural analyses which focus around a number of themes. First, political economy studies which examine the interplay between key agencies in formulating and implementing policies. Second, structural or regulation theory studies which often assume the primacy of capital in this interplay, or other structural factors which predetermine to a certain extent who will gain from policy changes.

Both of these approaches have the central concept that life is a struggle between people usually acting as organised groups;
iii) A human agency or behaviouralist approach however argues that human irrationality can undermine these structural forces, and that actors can act in different ways dependent on the issue at stake, thus farmers may pass up the chance to drain ponds and grow more crops because they are keen bird watchers. Similar approaches identify the importance of chance events, and factors such as chance have encouraged the development of a post-modern approach which rejects both the meta-narrative and structural approaches as arrogant simplifications of a chaotic complex universe.

The nature of our imperfect world also leads to a number of other key assertions in the book about the evolution of countryside planning policies:
a) Policies not only evolve via crisis driven incrementalism, but are added to in a process of accretionary incrementalism in which the policy matrix like Topsy, just keeps growing, since policies are only rarely junked;
b) Policies are affected by unexpected responses by those affected, and by unforeseen events;
c) Policies can be affected by policy developments in seemingly unrelated areas, notably general social policies, for example on education and housing, which can have major impacts;
d) Policies are created by individual departments or ministries and implemented in turn by discrete organisations. This leads to a fragmentation of policy and poor co-ordination between policies, as highlighted in point c) above; and
e) Policies are underresearched, in spite of the attempt since the mid 1980s to force public bodies to set out performance targets and develop systems of policy evaluation in a 'bottom-line' accountancy style of policy assessment used by the National Audit Office in their scathing attacks on the deemed inefficiency of many public policies. Research by academia and by pressure groups in contrast is often ad hoc, self-indulgent and focused on 'flavour of the month' topics, in spite of attempts by research grant awarding bodies to set the research agenda.

Thus the book adopts a broadly historical approach based on the key concept that people and the context within which they find themselves, make history. The book also adopts an organisational and sectoral approach, because policy is made by organisations which have been based on sectoral land uses, for example, agriculture, forestry, nature conservation, recreation and so on. Therefore, the book is based on three key approaches: historical evolution; organisational structures; and sectoral land uses.

The main aims of the book are also threefold. First, to provide an authoritative account and explanation of the evolution of countryside planning powers. Second, to examine different interpretations of the impact of planning policies on the rural environment, and third to take a long term perspective and assess how successful countryside planning has so far been in achieving its diverse aims.

An even longer perspective is provided by Cherry and Rogers (1996) in *Rural Change and Planning* which provides a comprehensive overview of the rise of planning in the first 70 years of the century followed by its subsequent metamorphosis in the last 30 years of the century. The book begins by lamenting, perhaps incorrectly, the tendency for modern academics to be a-historic in their outlook and unashamedly employs a synoptic narrative in order to tell a sequential story.

Accordingly nine of the book's eleven chapters cover specific periods. For example, chapters two, three, four and five cover the period till 1914, the inter war years, and the two world war periods. These chapters embrace all aspects of the countryside and take up about one-third of the book. After the Second World War however, the whole post war period is covered but in different sectoral chapters. This leads to a certain amount of overlap and repetition because the sectors are chosen by issues rather than by planning policy areas.

In more detail, the narrative chapters begin with the selection of five trends that are claimed to structure the century and that were already in place by 1914, namely: the decline of landed estates; nostalgia for a rural past; an anti-urban ethos; the adoption of middle class values in conservation; and the consequences of technological changes in road transport. Chapter Three explains how these trends were interrupted by the First World War before Chapter Four analyses how they became major themes in the inter war years, notably the dramatic decline in landed estates and agricultural production and the rapid growth of middle class suburbia.

Chapter Five argues that the Second World War was the midwife to the post war explosion of planning powers in that it provided a *focus* and added an *impetus*. It also argues that the ruling classes used their influence to ensure that planning would protect or enhance their interests, albeit under the guise of other aims, for example, producing badly needed food.

Turning to the sectoral post war chapters, Chapter Six examines agriculture and forestry in a breathless 18 pages which also includes agri-environment policies. Curiously, employment policies in Chapter Seven cover 14 pages even though it is acknowledged that these policies are strongly related to urban and global trends. Chapter Eight makes the common error of linking recreation and landscape protection policies and suffers also from a good deal of internal repetition as well as repetition with other chapters, notably with regard to Environmentally Sensitive Areas. Repetition is also a problem with the last two sectoral chapters which deal with communities and town and country planning.

All these detailed chapters are clearly written, albeit in a rather chatty style, and they are impressively comprehensive. This does however mean that detail is often sacrificed, except for a number of case studies or examples of policy development which go into greater depth. Readers wishing to follow up the material will however be disappointed since the references are sporadic and eclectic, and often seem to be chosen because they allow a catchy phrase like Shoard's the '*Theft of the Countryside*' to be woven into the text. This means that many illustrious names are absent or barely mentioned. In spite of these criticisms the book provides an admirably sound framework for anyone wishing to embark on a more detailed study of 20th century rural planning.

The true value of the book, as an analytical rather than descriptive text, must however be judged in terms of the themes it discerns in the first and last chapters. As already pointed out these are not cross referenced nor are they consistent. Thus in Chapter One the selected themes are: the economic drive of agrarian capitalism; the impact of urban values on the countryside; the crucial significance of agriculture for the landscape and thus the inseparability of rural planning and agriculture (why then does agriculture AND forestry only get one specific chapter?); the deep attachment to rural living; the conflict between development and protectionist measures; the rise of planning as an antidote to chaos and its decline as a collectivist egalitarian force in face of individualism and freedom.

In contrast, the final chapter selects five themes: protectionist versus development pressures; the rise of a multi-focus countryside; integration into a wider world; the growth of pressure groups who feel they have a 'stake' in the countryside; and the rise of experts and professionals in rural policy. Only one of these repeats the earlier themes, some are contradictory, and others are new themes.

This lack of overall conceptual structure reflects the descriptive nature of the book which is virtually theory free and ignores many of the debates of the last 20 years. Instead it employs an historical narrative and takes an incremental view because '*the story is one of continual adjustment of policy to changing circumstances, marked by the interplay of political, professional and special interest group pressure*' (p. 126). The book thus employs a personality dominated view and the interpretation and classification of events may be queried by certain commentators.

More constructively, the book provides an essential framework which can be used to construct more detailed and theoretically based interpretations of the century. In addition, the book correctly claims that planning has become an accepted process but one that has lost its purpose and vision in the last 20 years. The finest epitaph to this book and to Gordon Cherry who sadly died soon after it was completed would be to resurrect the vision of a better tomorrow which drove the founders of planning forward at the turn of the century. (Part of this review has already appeared in *The Journal of Rural Studies*, 14(2), 1998, pp. 267-8.)

Curry and Owen (1996) have edited *Changing Rural Policy in Britain*. This is a useful book which gathers together a collection of chapters which can be divided into those based on original research and those based on synthesis and policy analysis.

The research chapters contain: an examination of new housing and local planning policies in the Lake District by Gwyndaf Williams; a study of Rural Training and Enterprise Councils by Trevor Hart; an analysis of Conservation Amenity and Recreation Trusts by Ian

Hodge; an insider view of decision making by members of a local authority planning committee by Andrew Gilg and Mike Kelly; an exploration and analysis of environmentally friendly farming in South West England by Martin Battershill and Andrew Gilg; a study of how farmers perceive agri-environmental schemes by Helen McHenry; and a study of farmers' attitudes to woodland planting grants by Tim Lloyd, Charles Watkins and Daniel Williams.

The synthesis and policy analysis chapters include: an essay on changing rural policy by the editors; an assessment of the potential for landscape ecological planning by Paul Selman; an analysis of local authority policies on village planning and design by Stephen Owen; an essay on the sectoral division of rural planning agencies by Alexander Mather; a commentary on the evolution of CAP agri-environmental policies by Clive Potter; a synthesis of farm diversification and the environment by Bill Slee; and a synthesis of changing policy prescriptions in Environmentally Sensitive Areas with special regard to the built environment by Peter Gaskell and Michael Tanner.

The two sets of chapters thus provide a complementary collection, but as with all edited texts there are obviously some gaps, and some repetition, although to some extent this highlights where the current strengths and weaknesses of research into rural policy and planning are to be found. The main themes of the volume are the continuing division of rural planning themes into separate boxes and the incremental nature of policy making. Another major theme is the considerable uncertainty that surrounds the rural policy agenda in the mid 1990s as the post war consensus continues to break down. This volume thus represents a very useful position statement concerning current research trends and policy analyses, not only for students seeking a broad overview from a number of different perspectives, but also to future readers who will find here a useful reflection of where rural planning stood in mid decade.

Mackay (1995) has written an account of three of *Scotland's Rural Land Use Agencies* from the inside perspective of someone who worked for the Scottish Office for over thirty years. According to Moffat (1997) his analysis of the work of the Forestry Commission, the Nature Conservancy Council and the Countryside Commission for Scotland concludes that there is little hope of the fundamental issues in Scottish land use being tackled separately until they are tackled comprehensively.

Finally, Brown (1996) has edited the *Proceedings of the 1996 rural practice research Conference* which contains *inter alia* papers by Winter on agricultural science policy, Errington on small business growth, Lewis on woodland management, Cowap on Nitrate Sensitive Areas and Jones and Carver on the market for milk quotas.

2 The Environment

Environmental Legislation and Environmental Agencies

International and General Issues

Internationally, the UK continued to be in the vanguard of environmental policy and according to an OECD report (1994) the UK was first or second out of the G7 countries on ten out of 37 environmental indicators and third or fourth on a further dozen.

The Department of the Environment (1995a) issued an interim report on how it was converting the European Commission's Fifth Environmental Action Plan into a statement of domestic action, but while this was welcomed by pressure groups as a pioneering lead that other European countries should follow, they also called for full implementation of the Plan.

Within the UK the Government (Cm 3323, 1996) issued Europe's first environmental health action plan based on three groups of actions: securing basic environmental health requirements; seeking to prevent medium to long term environmental health hazards; and promoting well-being as opposed to the prevention of disease. The plan will be monitored in the annual environment White Paper. The Environment Department (1995b and 1996a) and the Transport Department (1996) also published reports on *Risk assessment and risk management for environmental protection*, *Environmental responsibility* and *The valuation of environmental externalities*.

The Government also published three more annual reports in the series which report on progress on the commitments made in the 1990 *This Common Inheritance* White Paper and the *1994 Sustainability Strategy*. In Cm 2822 (1995) the Government noted that it had created 16 more ESAs, set up several agri-environment schemes and imposed environmental conditions on CAP livestock schemes. The report also noted the Quality in Town and Country initiative, setting up machinery to take forward sustainable development. Cm 3188 (1996) noted the publication of the rural White Papers (see Chapter One) and procedures to create Environmental Agencies as a result of the passage of the *Environment Act 1995*. It also included new priority issues for 1996, for example, plans to put sustainability into practice by publishing an 'Environmental Health Action Plan for the UK'. Cm 3556 (1997) introduced a new emphasis for environmental policy, which for the previous 25 years had concentrated on cleaning up past pollution and improving standards. The new challenge is to tackle longer term issues of sustainable development which according to the report will call for a considerable culture change in our present lifestyle, one more attuned to conservation and the reuse of resources. As part of this culture change the report noted that 120 indicators for sustainable development had been published and that a priority for 1997 would be a review of how to environmentally appraise government policies. Finally, all three reports were dominated as before by lists of past commitments, action during the year and commitments to new or further action for a massive range of policy areas. The 1995 report was scrutinised via a House of Commons Environment Committee (1995a) which cross examined the Secretary of State for the Environment and two of his officials.

The Secretary of State for Wales (1995) issued an *Environmental Agenda for Wales* with six key elements: new planning guidance for local authorities, encouraging re-use of land and buildings and the regeneration of town centres rather than out-of-town retailing; more forestry to benefit local communities and nature; effective management of protected areas; more targeting of development on reclaimed land to reduce pressure on greenfield sites; further action to encourage more use of rail transport; and better use of resources through recycling, energy efficiency and other waste reduction measures.

Sustainable Development

The Department of the Environment (1996b) published a list of some 120 indicators of Sustainable Development in order to meet one of its commitments made in its 1994 Strategy for Sustainable Development. One third showed improvements over a 25 year period, one third showed a deterioration, with the rest being neither good nor bad. In more detail, people were travelling twice as far as they were in 1970 and consuming more than twice as much fuel. However, energy use had remained the same as 1970 in spite of a 60% rise in GDP, demonstrating that energy growth can be uncoupled from economic growth. The urbanised area stood at 10% in 1981 and was projected to rise to 12% by 2016. However, out-of-town floorspace had tripled between 1986 and 1990 and the average commuting journey had increased by 40% since the mid 1970s. Wildlife indicators showed that 25% of native species are threatened or scarce, and 22 species of farmland birds had declined over the last 20 years.

The Indicators were broadly welcomed as a first step, but criticised by some (Mullaney and Pinfield, 1996; and Levett, 1996) for not taking enough account of social equity and quality of life indicators while Bowers (1996) in a much more critical attack accused the list of being created by a committee scratching around among available and unrepresentative data to meet the needs of the list. Instead Bowers argued that the starting point should have been a discussion of what we need to know, followed by an exercise in determining the statistics that have to be generated and how this could be done. This theme had earlier been taken up by the Royal Society for the Protection of Birds (1994) which looked at the suitability of a number of indicators, including: soil quality; hedgerows; and woodland loss. They concluded that existing indicators failed because there were serious doubts about the accuracy of the data on which they were based and/or because the data were collected too infrequently. The World Wide Fund for Nature (1994) however felt confident enough to publish a list of 46 environmental indicators or 'green gauges'. These showed (Milne, 1994) that more than half the UK's hedges had vanished since 1947, and that more than 2,500 miles of rivers were graded as poor or bad.

In February 1995 the Government launched the 'Going for Green' initiative which aimed to stimulate and increase people's awareness in sustainable development, so that they would understand the real consequences of the choices they make as consumers, at work and at home, and thus hopefully change their personal and community life styles. In more detail Department of the Environment Circular 2/95

(1995c) informed local authorities that they could voluntarily join an eco-management and audit scheme as from 12 April 1995. Membership of the scheme requires local authorities to set out a policy which states their overall environmental aims, reviews the environmental impacts of their activities, sets out a programme of activities to achieve defined objectives, sets up a management programme for implementing the programme, provides for periodic audits to ensure that the programme is being followed, and a statement of environmental performance with impartial, external verification of the process. The Local Government Management Board (1995a) published a pragmatic guide (Milne, 1996a) on how to use around 20 indicators in developing Local Agenda 21 programmes based on the experience gained by a group of ten local authorities who had piloted the indicators.

A series of recommendations on sustainable development issues was provided in three reports by the British Government Panel on Sustainable Development (1995, 1996 and 1997). For example, the first report (1995) urged the Government to give higher priority to defining its environmental objectives. The second report (1996) continued this theme and called for more decisive action in the setting of objectives with transport and agriculture seen as priority areas. In addition, it recommended a national forestry strategy containing incentive measures for economic, environmental and social goals and identifying areas where forestry expansion could take place. The third report (1997) identified four issues for higher priority. First, introducing environmental criteria into Government procurement procedures. Second, a review of all Government subsidies in terms of their environmental impact with a view to reducing their impact. Third, to develop a strategic energy policy. Fourth, a fresh initiative is needed to reform the CAP, to redirect funds from agricultural commodity support to direct environmental payments, and to make sustainable farming a central objective.

The UK Round Table on Sustainable Development which was set up in 1995 also published a number of reports on Sustainable Development, for example, on freshwater, urban transport, and on 'Housing and Urban Capacity' which tested the proposition that 75% of new housing should be built on previously developed land. Its aim is to encourage discussion on major sustainable development issues and to build consensus between different perspectives and different responsibilities. It has about 30 members drawn from central and local government, business, environmental organisations and other sectors of the community. Although co-chaired by the Secretary of State for the Environment this does not imply endorsement of the Round Table's recommendations.

A final set of recommendations was provided by a major report from the House of Lords Select Committee on Sustainable Development (1995) which examined the Government's strategy as set out in Cm 2426 (1994). The report was backed up by a mountain of evidence from, and cross examinations of, the leading thinkers and organisations in the field. The report argued that sustainable development implies a revision of the path of wealth creation and constraining parameters of economic decision-making by a recognition of the environmental costs of development. In essence it called for a higher priority for sustainable development in government thinking and policy and the need for precisely specified environmental targets to be set to which the Government responded in Cm 3018 (1995). The main recommendations and the responses are now dealt with in turn.

Both parties agreed that the definition of sustainable development should be a pragmatic one. Turning to scientific uncertainty and the precautionary principle the Committee argued that they should not be overused and veered between the wait and see approach and the need to take action based on political judgements even before the science was clear, an equivocation which the Government welcomed. The Committee was however clearer on the need to set targets and the Government confirmed that they would be ready to do so when this would be sensible. With regard to specific targets both parties agreed on the need to reduce traffic, the use of cars and to reform the CAP. The Government did not agree with the Committee's view that national forestry targets should not be set.

Turning to policy measures both parties agreed that economic instruments had the potential to become more widely used but that regulation would always be an important instrument, with the Government more enthusiastic about the use of cross-compliance to secure environmental gain as in the case of reducing stoking rates in livestock farming. The Government agreed with the Committee about the difficulty of costing sustainable policies and recommended the greater use of cost-benefit analyses. The Committee expressed concern about the integration of environmental policies and between the UK and the EU and in their response the Government resolved to work for closer integration. Finally, both parties stressed the need for lifestyles and attitudes to change and thus the key role of education and appropriate fiscal and legal measures in raising environmental awareness.

A number of books were also published demonstrating how sustainable development could and should be put into practice. For example, Selman (1996) provides a very good checklist of principles and methods logically set out with good examples. It does however represent a rather mechanistic systems view of planning processes, albeit modified by consensus building and participation exercises. It argues that sustainability is now on the permanent agenda of decision makers but implementing it poses a challenge for democracy. On the one hand green activists may dominate the process and fail to gain a broad basis for support, on the other hand, representative democracy may fail to inspire citizens of the need to follow sustainable strategies. Thus, if local sustainability strategies are to become genuinely sustainable, they must somehow combine the legitimation of local government with the vigour of community action. Selman concludes that this may take a generation for the necessary changes in attitudinal and behavioural action to occur.

Three further books by Hams *et al* (1994), Kirkby *et al* (1995) and Ageyman and Evans (1994) also provide examples of how sustainable development policies could be put into practice at the local level, for according to Pinfield (1994) it is in local authorities that the innovation, dynamism and commitment to deliver sustainable development lies. Accordingly, the Association of Metropolitan Authorities (1996a) *Environmental Manifesto for Local Authorities* which sets out the existing policy objectives of the Association and a series of new policy recommendations is particularly welcome.

The 1995 Environment Act

The Act began as a House of Lords Bill (HL 10(94-95)) in December 1994. Its main purpose was to establish an Environment Agency for England and a Scottish Environment Protection Agency. It also included measures: to establish independent National Park Authorities (later in this chapter); to protect important hedgerows (see Chapter Four); and to transfer the Countryside Stewardship Scheme from the Countryside Commission to the agricultural ministries (see Chapter Four). The rest of this section thus concentrates on the Environment Agencies.

During the passage of the Bill the main points of dispute centred on the duties of the Agencies and whether they should consider not just the benefits of environmental protection measures but also their costs. Burton (1995) has argued that many concessions were won, but at the same time some difficult truths were confirmed about the place of the environment in the then Government's priorities. However, the centrepiece of the Act when it received the Royal Assent in July 1995 was the creation of the Environment Agency which took over the duties of the National Rivers Authority, Her Majesty's Inspectorate of Pollution, the Waste Regulation Authorities and other functions of central and local government relating to the environment as from 1st April 1996.

The principal aims and objectives of the Agency are to protect or enhance the environment and to make a contribution towards attaining the objective of achieving sustainable development in a manner which Ministers consider appropriate. In doing so it must take into account a number of environmental factors but also the effect any of its proposals would have on the economic and social well-being of local communities in rural areas. However, according to Burton (1995) much will depend on how the detailed regulations to implement the Act are written and interpreted.

The Scottish Environment Protection Agency also began work on 1st April 1996 with the general duty of preventing or minimising or remedying or mitigating the effects of pollution on the environment (Henderson, 1995). It has three specific duties: promoting the cleanliness of rivers, inland waters, ground water and tidal water; conserving water resources; and promoting the conservation and enhancement of the natural beauty and amenity of inland and coastal waters and the conservation of aquatic fauna and flora. However, it also has to have regard for the social and economic needs of Scotland, and in particular the rural areas.

The Environment Agency

The Environment Agency (1996a) published its first corporate plan for 1997-98 in November 1996 in line with the Environment Department's (1996c) statutory guidance. This included the following tasks: assess the state of the environment and the pressures placed upon it to help both Government and the agency determine the necessary management responses; develop a strategy which contributes towards the national objective of sustainable development taking into account risks, costs and benefits; protect and improve the environment by effective regulation, through its own actions and by working with and influencing others; develop a research programme so that decisions and actions are based on sound scientific and technical advice; and provide high quality information and develop an environmental education strategy. As part of the corporate plan the Environment Agency (1996b) began the creation of a new and comprehensive inter-active state of the environment database because past techniques may have given a distorted view.

The Department of the Environment (DOE)

The DOE underwent a far-reaching reform of its organisation and working methods in 1995 which reduced the number of management levels and in order to promote cohesion across the Department four new strategic aims were adopted: sustainable development; maintaining and improving the quality of the natural and built environment; achieving economic competitiveness; and good government.

The work of the Department was covered in a number of documents. The Annual Report for 1994-95 (Cm 2807, 1995) recorded that the DOE had: set up a Biodiversity Action Plan Steering Group; implemented the EC Habitats Directive; given the

New Forest equivalent planning protection to National Parks; commercially launched the Countryside Information System; published new data on the loss of hedgerows; launched the Rural Challenge competition; introduced new Rural Development Areas; published a Citizen's Charter Guide on the Planning System; and launched the 'Quality in Town and Country' initiative.

In a commentary on the report the House of Commons Environment Committee (1995b) remarked that the environmental protection budget of £221 million was very small at less than one per cent of total expenditure. However, evidence given to them had estimated that £15 billion was being spent by the private sector on environmental protection which encouraged them to believe that the 'polluter pays principle' was being successfully implemented.

The report for 1996 (Cm 3207, 1996) recorded that the DOE had: published the Rural White Paper jointly with MAFF; consulted on a list of 280 possible Special Areas of Conservation (SACs) of which 211 had been submitted to the European Commission; established the National Forest Company and launched the National Forest competitive tender scheme; approved business plans for the remaining nine Community Forests; and accepted the main recommendations of a joint Police/DOE report on enforcing wildlife laws.

In a commentary on the report the House of Commons Environment Committee (1996b) expressed concern that the Committee of Green Ministers was concentrating on internal housekeeping. Accordingly, the Committee asked for a fuller discussion of how the DOE was pursuing its mission of integrating environmental concerns across government.

This was addressed in the report for 1996-97 (Cm 3607, 1997) which provided a matrix of interlocking aims and objectives, including a new objective of promoting public health. In addition the report recorded that the DOE had: spent around £120 million on its countryside and wildlife work; completed the review of the local government structure of England which created 46 unitary authorities in total; published a Green Paper on options for meeting the projected 4.4 million increase in the number of households; made a significant contribution to the Transport Green Paper published by the Department of Transport; approved the National

Forest Corporate Plan; published *Water Resources* a strategy for the sustainable provision of water in the longer term; and overseen the successful second year of the National Forest Company tender scheme and the first full year of the twelve Community Forests.

The Countryside Commission

The annual report for 1994-95 (Countryside Commission, 1995b) began by expressing relief that the merger with English Nature had been called off in September 1994. Other events in the year were approval for the National Forest in the Midlands and the 12 Community Forests and the continued success and expansion of the Countryside Stewardship Scheme. The report for 1995-96 (Countryside Commission, 1996b) welcomed the adoption and expansion of the Countryside Stewardship Scheme by MAFF, even though this meant the loss of 57 staff from 318 to 261 and a budgetary cut from £42.7 to £25.7 million. The report also noted that 80 per cent of its targets had been met, including a successful bid to the Millennium Commission for funds for Millennium Village Greens, significant progress on improving rights of way and the development of a 10 year strategy for the Commission's work.

In *Working for the Countryside*, the Countryside Commission (1994a) outlined how it was achieving its corporate plan for 1994/95 to 1997/98 mainly by working with people and organisations within the basis of strong policies. Finally, a 12 page report by the Countryside Commission on its activities and a cross examination of the Chairman and the Chief Executive by the House of Commons Environment Committee (1997a) provides a useful insight into how the Commission operates.

Scottish Natural Heritage (also see Chapter Five)

Scottish Natural Heritage published a number of annual reports and corporate plans which showed that it had successfully merged its two former bodies, the Countryside Commission for Scotland and the Nature Conservancy Council in Scotland. It also published a major report on the state of Scotland's natural heritage (1995a), a document which challenged more people to demonstrate their care for the environment practically (1995b), and a document which set out its priorities (1996) which emphasised the need to work with all public bodies within a framework of biodiversity and environmental education, access and enjoyment.

The Countryside Council for Wales

Information about the work of the Council is provided in its annual reports and annual reports by the Welsh Office (Cm 2811, 1995; Cm 3211, 1996, Cm 3615, 1997) and also in the form of a cross examination of some of its members by the Welsh Affairs Committee (1996) which unfortunately neither contains any written evidence or a report by the Committee. These reveal that the Council faced severe cuts in 1995-96 with its budget cut from £20 to £17 million and that many of its roles were under critical scrutiny by the then Welsh Secretary, John Redwood (Spray, 1994). However, by 1996-97 its budget had been expanded to £22 million and both its future and its overall programme seemed secure. Looking to the future the Council (1996) published a discussion document on the future of the landscape which floated the prospect of AONB status for the Berwyn Mountains.

English Nature (Main Report in Chapter Five)

English Nature (1995a) set an example to other environmental bodies when it published an environmental strategy which included a plan to be 'green' in its own day to day activities but elsewhere in its work to focus on long term conservation and sustainability and rarely to involve itself with day to day issues.

Consortia of Agencies and Pressure Groups

In the autumn of 1995 the Countryside Movement was launched as a campaigning voice for country people to counter threats to their way of life. It was supported by the NFU, CLA and the main field sports groups, but not by the Ramblers and the RSPB. However, it was dissolved in the spring of 1997 when it ran out of cash. According to Tayhoe (1997) it failed because although it attempted to provide an integrated view of rural life and to take an holistic view which could represent the views of all rural groups it was identified as a field sports lobby and did not appeal to most people interested in rural issues who now ironically live in urban areas.

A more formal attempt at an holistic view was provided by the Countryside Commission, English Nature and English Heritage (1996) who published a document which showed how some local communities had already embraced Local Agenda 21 and how their example could be followed by others. In addition a report by Aston Business School (1997) reviewed the efficiency and effectiveness of the Rural Action for the Environment Initiative funded by the Countryside Commission, English Nature and the Rural Development Commission.

Pressure Groups

The Council for the Protection of Rural England (1996a) celebrated 70 years of achievement which have included safeguarding precious countryside from the first motorways, achieving environmental reforms to agricultural policy, stopping damaging tax breaks for conifer planting in the hills in 1988, and resisting the case for new settlements in the 1980s.

Friends of the Earth (1994a) published a report which claimed that more stringent environment policies could help cut pollution, conserve resources and create employment for 700,000 people and thus save £3 billion in unemployment. In more detail: if one-quarter of farming went organic agricultural employment would rise by 45,000; making polluters pay for clean ups could create 200,000 jobs; and if 600,000 hectares were planted for timber production that 3,300 to 4,000 jobs would be created.

Water Issues

The Department of the Environment (1995d) published an action plan, during one of the driest summers for 250 years, which confirmed the Government's view that the maximum use should be made of existing resources. This approach was however criticised for ignoring the potential of land-use planning in directing new development towards areas with adequate water supplies and for not setting mandatory leakage targets for the water companies, which would have been provided by a failed private members Bill, the *Water (Conservation and Miscellaneous Provisions) Bill 1995* (HC 185 (94-95)). The Department of the Environment (1996d) in a longer term perspective, *Water Resources*, then according to Milne (1996b) passed the buck on long term water planning to both the Environment Agency and the water companies. The Environment Agency was picked out as the body who should plan water resources at the national and the local level, while, the water companies would be expected to prove via the local planning system and the water regulatory system that there was no alternative to meeting projected demand other than by creating new resources, for example, up to seven major reservoirs in the south east of England.

The House of Commons Environment Committee (1996c) in two reports criticised the confused division of responsibility between the DOE and the two regulators and the water companies for not knowing how much water it needs nor how much is being lost in the supply system. The Committee thus called for a big drive to reduce waste use and more sophisticated appraisal of existing reservoir capacity and groundwater storage. Only when this had been done could there be a case for new reservoirs, and then only after a strategic environmental assessment had shown that existing resources had been used to the full, and that all environmentally less damaging options, including demand management, had been tried.

In their response, the Government (Cm 3562, 1997) welcomed the report since it broadly mirrored the views expressed in the Department of the Environment (1996d) *Water Resources* publication outlined above, and indeed most of the responses drew widely from the 1996 document. In more detail five themes were of interest to rural water issues. First, both documents agreed that the top priority is to cut leakage in the system by 15%. Second, agricultural irrigation only accounts for about one per cent of water usage, but nevertheless this should be based in future on farm reservoirs which would store winter rain. The Government agreed to revise PGG7 to make the construction of such reservoirs acceptable to local planners. Third, the Government accepted that forestry developments can have an impact on hydrology and thus they agreed that possible impacts would be taken into account when determining the type and location of new forestry development. Fourth, the Committee argued that there could be a case for a 'Silkin' test to apply when evaluating the case for extensive new water schemes. The Government responded however, that this would be inappropriate since the Water Act 1991 contained enough environmental safeguards. Fifth, the Committee wanted a long term strategy for water resource planning to be created but the Government argued that this was to some extent already provided by the monitoring work of the Environment Agency carrying on from the National Rivers Authority and that anyway water was best planned on a system by system basis from which sustainable regional and national strategy can be built and put into action.

The Council for the Protection for Rural England (1995a) also endorsed the more efficient use of water and reducing demand, and via experts, Newson *et al* (1996) they also called for better co-ordination between the Environment Agency and local authority planners in linking developments with water resources. In contrast to the water industries demands for new reservoirs the Institute of Terrestrial Ecology (1995) found that the number of ponds had probably fallen by 9% in the latter half of the 1980s.

Holistic Environmental Areas

National Parks

The *Environment Act 1995* revised and updated the purposes of National Parks as set out in the 1949 National Parks Act to be one:

'*(a) of conserving and enhancing the natural beauty, wildlife and cultural heritage of the areas.....; and*

(b) of promoting opportunities for the understanding and enjoyment of the special qualities of these areas by the public.' (Section 61)

The Act also introduced two new duties for National Park authorities: first the authorities should seek '*to foster the economic and social well-being of local communities within the National Park*'; and second, all public bodies should have regard to National Park purposes except when there appears to be a conflict between these purposes, in which case they should '*attach greater weight to the purpose of enhancing the natural beauty, wildlife and cultural heritage of the area*' (Section 62).

The revised and new purposes had been hotly debated during the passage of the Act from its introduction as a Bill in late 1995 to its royal assent in the summer of 1996. Most of the debate centred on whether the word 'quiet' should be used to describe enjoyment as advocated by the 1991 Edwards report. The House of Lords inserted the word 'quiet' in February 1995 but it was deleted at the final reading stage in June 1995 even though it had support from almost all the rural conservation pressure groups.

Some critics were also disappointed that the Act had not contained a statutory test for proposed major development in the Parks to give greater weight to the 'Silkin test' but this was rejected by the Government on the grounds that the Government already had a policy that major developments should not take place in the National Parks, save in exceptional circumstances.

The Act also wound up the mixed system of National Park Committees and Boards and replaced them with new National Park Authorities who became the sole local planning authority (including minerals planning) with the additional duty to prepare a National Park Management Plan within three years of coming into operation unless such a plan had recently been approved by their predecessors. These Plans should be reviewed every five years. The new authorities took over power on 1st April 1997 and became independent of county council control. Funding however remained the same, with 75% coming from central government, and 25% from the new authorities.

Advice on implementing the Act was provided by a Department of the Environment (1996e) circular which noted that the Sandford principle of endorsing conservation as a last resort was now enshrined in Section 62. The Circular also spelt out in detail how the members should be appointed. Namely, one half plus one must be local authority appointees, while the remainder should be appointed by the Secretary of State for the Environment with one half minus one being drawn from parish councils in the area and the rest being national appointees. The Act thus diluted the presence of national appointees in order to bring in new parish representatives. The

Circular also dealt with the five types of plan to be prepared by the authorities: the Structure Plan; Local Plans; minerals plans; waste local plans; and a management plan. Finally, Circular 4/76 was replaced, but the new circular noted that the Annex of 4/76, which discussed the Sandford Report, remained of interest.

Character Map of England

English Nature and the Countryside Commission (1996) with help from English Heritage produced a map of England which depicts the natural and cultural dimensions of the landscape. The map is intended to be used for a variety of purposes including: identifying where and how agri-environment schemes might best be extended and tailored to specific parts of the country; and helping in decisions about the siting and composition of new woodland, with a view to achieving the target of doubling tree cover by 2050 in ways that respect local character. The map depicts 181 character areas for use by the Countryside Commission and 159 Natural Areas (some of which are aggregated character areas) which vary in size from the Isle of Portland to the Fens.

Areas of Outstanding Natural Beauty

The Tamar Valley, straddling the Devon and Cornwall border, became the 41st and probably the last AONB in England and Wales when it was designated in August 1995 (Countryside Commission, 1995c). The Commission also published landscape assessments for a number of existing AONBS: the High Weald (1994b); Cannock Chase (1994c and 1995d); the Cranborne Chase and West Wiltshire Downs (1995e); the Solway Coast (1995f); the Kent Downs (1995g); the Norfolk Coast Landscape (1996c); the Wye Valley Landscape (1996d); the Howardian Hills (1996e). The Countryside Commission (1996f) also proposed that the Sussex Downs and East Hampshire AONB should be overseen by a new conservation body covering the whole of the South Downs.

Miscellaneous Issues: The Defence Estate; the Cairngorms and Heritage Coasts

The Countryside Commission (1994d) gave advice to the Ministry of Defence about the proposal to develop the Otterburn training range in the Northumberland National Park. This proposed that no action be taken unless it was fully justified and met the tests set out in PPG7, and that in the wider perspective and for the medium term the Ministry of Defence should participate in an independent national review of military training needs as recommended by the 1991 Edwards report. This was endorsed by the House of Commons Defence Committee (1995) which advocated that the DOE and the Ministry of Defence should provide general policy guidance on the reuse of defence sites expected to be released from military use as a result of the peace dividend, notably in green belts, but that there was little prospect of defence land being released in the National Parks or even cessation of training activities.

In Scotland the Cairngorms Partnership Board was set up in November 1994 (Burton, 1994) with a view to developing an integrated approach to the problems facing the fragile mountain area by bringing together the powers, funds and expertise of the partner bodies to ensure that the mountains are sensitively managed. Critics of the Board argued that it would lack power to prevent intensive agriculture and tourism in the area and the fact that it had not been given statutory consultee powers for planning applications (Reid, 1995).

The Countryside Commission (1995h) updated 1992 advice on Heritage Coasts and set out targets for the year 2000 which emphasised their twin role as catalysts and watchdogs, the potential to make more economic use of their quality environment, and the key role they could play in Local Agenda 21 programmes.

Preserving the Built Heritage

The *National Heritage Act 1997* amended the National Heritage Act 1980 by extending financial assistance for projects which relate to the national heritage or the history, natural history and landscape of the UK.

In England, English Heritage (1996) published documents on archaeology and nature conservation in wetlands and with the Countryside Commission and English Nature, English Heritage (1996) also published an advisory document which urged local authority planners to fundamentally reassess the relationship between the natural and the man-made environment when preparing local plans. This argued that conservation and development are not alternatives, but are contrasting interests that must be reconciled and integrated within the concept of sustainable development.

Planners were also provided with conservation advice in an updated PPG15 (Environment, 1994a) which placed greater stress on keeping historic buildings in active use as the best way of preserving them, but reminded them that keeping areas alive and prosperous would also mean not imposing unnecessarily detailed controls over business and householders. Two years later a consultation paper from the Department of National Heritage (1996) considered the case for more consultation over listing procedures, dropping Crown exemption, and procedures for listing post war buildings. Earlier, in 1995, the Department of National Heritage had issued advice on how reorganised local authorities should deal with conservation. This argued that the bigger authorities should set up specialist teams, and that smaller authorities should enter into management agreements with another authority to ensure that the necessary expert advice was available.

A review of PPG16 *Archaeology and Planning* by Roger Tym (1995) for English Heritage found that local planning authorities are giving archaeology issues the attention they deserve in dealing with planning applications, but that they could make wider use of Article 4 directions to protect areas of archaeological potential.

The House of Commons Committee of Public Accounts (1996) expressed concern at the slow progress made in Scotland on the resurvey of listed buildings and the schedule of ancient monuments especially since the current list is obsolete. At the same time a consultation paper (Cm 3267, 1996) reported that the listed building system was working well, so that only fine tuning changes were recommended, notably with regard to the listing of post war buildings. The paper also considered the merits of removing crown immunity from listed building control. No changes were made however in the *Planning (Listed Buildings and Conservation Areas) (Scotland) Act 1997* which consolidated all legislation involved with the built heritage into one Act.

In Wales the House of Commons Welsh Affairs Committee (1997) in a follow-up report to a 1993 report congratulated CADW on accelerating their resurvey of listed buildings, while stressing the need to get on with the work, especially with regard to chapels. The Committee also urged CADW to be more proactive in support for the integrated protection of wider historic landscapes in Wales.

Surveys, Management and Design, Climate Change and Commentaries

Surveys

The Department of the Environment (1994b) published five more volumes in its 'Countryside 1990 Series' which examine land use change and included a volume on the development of the Countryside Information System as an attempt to provide an integrated approach to the data handling needed to support rural policy decisions. The Countryside Information System is run by the Institute of Terrestrial Ecology and provides a mass of data at the kilometre square level either on the Web or via floppy disks. In Scotland the Scottish Office (1996a) considered the issues involved in land cover projects.

A survey of tranquillity by the Council for the Protection of Rural England (1995b) demonstrated a dramatic erosion in tranquil areas from 70% of England in the 1960s to only 56% in the 1990s. The worst erosion had occurred in the South East down from 58 to 38%, while the former most tranquil area, the South West, fell from 83 to 66%, leaving it second to the North East at 68%.

Management and Design

The Institute of Environmental Assessment and the Landscape Institute (1995a and 1995b) produced guidelines on landscape and ecological assessment which according to Gill (1996a) should become the two standard sources for planners. The guidelines urge practitioners to differentiate between quantitative measures, based on the magnitude of change, and more qualitative measures based on the significance of change.

The Countryside Commission which had supported the previous publications made their own contribution in two publications focused on design. In the first, the Commission (1994e) demonstrated how experiments in design statements had shown how they could help retain local character, while in the second, the Commission (1996g) used this experience to set out how local communities could participate by creating their own Village Design Statements.

In Scotland, Wightman (1996) on behalf of Scottish Wildlife and Countryside Link has provided according to Gill (1996b) an incisive analysis of the mismanagement and inappropriate development of Scotland's montane habitats. The report then highlights opportunities for more controlled and integrated land use, an expansion of public ownership, enhanced participation by and benefits to mountain communities, and the restoration of native forest and montane heaths and scrub.

Climate Change

The United Kingdom Climate Change Impacts Review Group (1996) predicted warmer temperatures with more rain in the north but less in the south between 2030 and 2050. This would have positive benefits for forestry and would allow new agricultural crop varieties to be introduced. Similar predictions were made by a Countryside Commission (1995i) review which suggested a northward shift of 300 kilometres and an upward shift of 200 metres with greater rainfall in the south offset by greater evaporation leading to more summer droughts and thus more demands for water (Environment, 1996f).

Commentaries

Myerson and Rydin's book on *The Language of Environment* (1996) is an attempt to understand the cultural presence of the environment by actively engaging with environmental debates culled from a wide variety of texts. These were then subjected to analysis to provide the three main themes of the book. First, that texts on the environment should be viewed as rhetoric in the hope that this will help: to understand the dynamic and creative processes of argument in a potentially democratic society; to search for plurality; and to accept difference and contradiction. Second, that the ethos adopted, is probably the central factor in environmental arguments, since truth is not claimed in a vacuum, but from a point of view. Thus we need to know how truth is claimed, on whose behalf, and with what methods, assumptions and perspectives? Third, that knowledge will not settle environmental disputes by itself, but that in common with the views of Giddens and Lindblom the explosion of knowledge could lead to a new plural enlightenment and a participatory democracy whose principle would be open-ended discussion.

This book is a well-written and constructed attempt to help people read more intelligently about the environment, and it certainly presents familiar material in an innovative way. However, it isn't really what it claims to be, about language, but is in fact a book about environmental arguments and in that sense it may fail to attract readers who may fear a terrifying text on semiotics. This would be a pity since this book clearly owes a great debt to the lucid writings of Tim O'Riordan over the years, and it should be read in parallel with his spectrum of environmental ideologies. However, it is not really clear to who the book is directed, since it is neither a classical student text nor a research report. Instead it combines elements of both. Perhaps it is directed to everybody since it concludes that the environment should be the centrepiece of democracy. Unfortunately, the title may put people off.

Wilson and Bryant (1997) seek to contribute to a new understanding of environmental management by moving away from the traditional state-centric approach. In contrast it seeks a more 'inclusive' understanding about environmental management and managers designed to embrace all types of managers conceptualised as interacting in a multi-layered process.

It does this by setting out a number of classificatory typologies based on observations from around the world. For example, six types of environmental managers are identified under the headings of: the state; environmental NGOs; Trans-National Companies;

International financial institutions; farmers; and hunter-gatherers. Their individual actions and interactions are then considered under the broad headings of 'Uncertainty' and 'Predictability' before the book concludes with a discussion about the future which ends with a plea for environmental management to be seen as a core social science.

The book thus represents a brave, albeit over-rigid attempt to reorder existing ideas and knowledge into a set of matrices. As such it is old wine in a new bottle, rather than a new elixir. Nonetheless, the core idea of environmental management as an organising concept aligned to ideas of sustainability do perhaps pose new conceptual structures. However, in terms of achieving predictable rather than uncertain ecological developments in the real world, even the authors are rather pessimistic about the methods currently employed by environmental managers since they tend to produce conflict rather than consensus. In conclusion, an interesting way of rearranging concepts, but one that may have just as well been done in a long paper, rather than a book.

Other books which take an environmental overview were provided by Mather and Chapman (1995) on *Environmental Resources*, Barrow (1995) on *Developing the Environment*, Buckingham-Hatfield and Evans (1996) on *Environmental planning and sustainability*, and books on *Environmental Law* by Leeson (1995) Freshfields (1996) and McEldowney and McEldowney (1996).

3 Town and Country Planning

The Policy Context

Legislative Change

There were no major legislative changes in the period under review, but there were some minor changes to the *General Development Order* and the *Use Classes Order*. In particular, a 1997 amendment to the GDO attempted to plug a loophole with regard to agricultural buildings. Under the revised GDO, if a farm building is erected as 'permitted development' but ceases to be used as a farm building, it should be demolished, unless planning permission is granted for an alternative use within three years of the agricultural use ceasing. This was intended to plug the loophole, whereby farm buildings were erected without the need for planning permission, but were then used for another purpose for which planning permission would have been required. However, Harrison (1997) argued that the legislation will be difficult to interpret and enforce.

The *Local Government and Rating Act 1997* gave new powers for the Secretary of State for the Environment to require improved consultation by county and district councils with the parish tier on a range of matters including planning policy and its implementation. These new powers implemented proposals made in the Rural White Paper.

One proposed change that failed to materialise, was the 1994 proposal to allow individual local planning authorities to set planning application fees at a level which would enable them to recover the costs they incur in processing planning applications. In January 1995, the Government felt that not enough support was forthcoming to support the change and thus continued with a standard set of fees, which were increased by 15% on average, with the medium term aim that on average 100% of the cost of determining planning applications should be met through the fee system. A further increase of 10% in December 1996 brought that goal nearly to fruition.

In November 1996 the Government announced its intention to allow householders to extend their gardens into surrounding farmland, after a number of high profile cases had shown heartbroken people having to surrender their extended gardens back to farmland, because they had failed to seek planning permission.

Advice in PPGs, DOE Circulars and Related Documents

PPG1 General Policy and Principles (Environment, 1997a): was the third version of this important policy statement. It:

- reaffirmed the role of the planning system in meeting the needs of a growing and competitive economy, in providing for new development, such as housing, and in protecting the natural and built environment;
- restated the limited circumstances in which it is appropriate to use planning obligations to secure development;
- reaffirmed the Government's commitment to a plan-led system of land use planning; and
- in a new section on propriety, advised that elected members should make planning decisions on the basis of an officer's written report.

The new PPG1 also provided guidance on sustainable development and stressed the need to encourage different activities to locate close to each other in order to minimise travel and promote community spirit.

The 1997 PPG1 was thus very different from the original PPG1 in 1988 and its previous revision in 1992, in that the 1988 version didn't mention 'sustainable development' or 'mixed uses' at all. In the 1997 version they are used as positive virtues in favour of encouraging the development of brownfield sites and urbane lifestyles, so that the countryside can be protected. Another change is in the greater emphasis given to following policies in the development plan and the declining emphasis given in favour of development in line with Section 54A of the Town and Country

Planning Act 1990. The revised PPG1 and a very similar draft, issued in July 1996 were broadly welcomed by environmental groups (Delafons, 1996 and Carmona, 1996) as a step in the right direction, albeit a step that could have gone further down the environmental road.

PPG7 The Countryside-Environmental Quality and Economic and Social Development (Environment, 1997b): revised and replaced the 1992 version of PPG7. The main changes were to:

- take account of the White Paper *Rural England*, and of PPGs published since 1992;
- advise on achieving good quality development and respecting the character of the countryside;
- re-state and clarify policy on protecting the best agricultural land;
- clarify policy on the re-use of rural buildings, allow greater discrimination in favour of the re-use for business rather than residential purposes, and advise on incorporating a residential element within a scheme for business re-use;
- stress the importance of thoroughly checking the lawfulness of developments to be carried out under agricultural permitted development rights, and advise on the possible removal of new buildings erected under them but not used for agriculture;
- strengthen the agricultural dwellings concession to counter abuse; and
- advise on local countryside designations and on the planning implications of Rural Development Areas and European Union Objective 5(b) areas.

The PPG broadly followed the draft version which was issued in the summer of 1996. The majority of the 400 plus responses had felt that it struck a fair and realistic balance between environmental protection and the need to stimulate economic activity. Indeed the Rural Development Commission welcomed the PPG in that it recognised that healthy economic activity helps to protect and improve the countryside. Support for the PPG also came from the CLA, the NFU and the CPRE although all of them expressed concern over some aspects of the advice, with predictably the farming lobby questioning the strengthening of controls, while the CPRE argued for planning controls to be further strengthened. The RTPI and the Countryside Commission also welcomed PPG, but both argued that it could lead to more rural traffic as a result of some new housing and employment being encouraged. Because the Government had not made any progress on the White Paper commitment to introduce a separate use class for rural businesses (Johnston, 1997a) it was announced at the launch of PPG7 that a working party would be looking into the possible introduction of a new use class. Finally, one commentator (Perraton, 19976) argued that the advice was only a very pale shade of green and that it only tinkered at the margins of sustainability, while Fairlie (1996a) argued that the new advice would still not prevent abuse of the agricultural loopholes by wealthy people, but would still prevent low impact smallholders from setting up sustainable organic farms.

PPG9 Nature Conservation (Environment, 1994c): advised on the planning implications of the regulations which implemented the EU Habitats Directive which came into force on 30 October 1994. The main feature of these regulations is that the GDO no longer provided permitted development rights for developments which would adversely affect the integrity of an area designated under the Birds or Habitats Directive. The PPG also gave advice about how to integrate nature with economic development outside nature conservation sites, and advised that where conflict is unavoidable the adverse effects on wildlife must be minimised. The PPG, which was the first advice since DOE Circular 27/87, was broadly welcomed (Culley, 1995 and Tyldesley, 1994) but some people argued that some of the advice would be a lawyer's dream, for example paragraph 27 which reads: '*local authorities should not refuse permission if development can be subject to conditions that will prevent damaging impacts on wildlife habitats or important features, or if other material factors are sufficient to override nature conservation considerations*' (Taylor, 1994). Furthermore, some local authorities and staff in conservation agencies claimed that this paragraph would force local authorities to weaken existing local policies. Nonetheless, by emphasising the need for conservation not just in the hierarchy of protected sites but everywhere the PPG marked a major step forward for nature conservation.

PPG2 Green Belts (Environment, 1995e): made limited changes to the previous PPG in light of the comments received on a draft revision issued in 1994. Six main amendments were made:

1) Reaffirmed the Government's commitment to Green Belt policy, stressed the fundamental aim of preventing urban sprawl by keeping land permanently open;
2) Set out positive objectives for land use, so as to secure greater benefits for the environment without compromising the overall restrictive approach;
3) Further encouraged proper consideration of the long-term direction of development, through the development plan system;
4) Removed the concession for new building by institutions;
5) Enabled local planning authorities to make realistic provision in their development plans for the future of existing major developed sites in the Green Belt, such as pre-war factories and power stations, in such a way to secure environmental benefits; and
6) Introduced a new policy on the re-use of buildings.

Limited infilling is still allowed and the PPG introduced a new type of limited infilling for community needs in line with PPG3. One area that was left in doubt however (Pitt, 1995) was the acceptability of giving permission for a development, the profits from which would 'enable' an environmentally beneficial development, notably, the creation of a community forest in the Green Belt. Another area of doubt concerned development which would improve damaged or derelict land (Taylor, 1995). Overall, however, the revised PPG was welcomed as a further strengthening of Green Belt policy.

PPG6 Town Centres and Retail Developments (Environment, 1996g): signalled the end of the out-of-town shopping centre boom of the 1980s. It argued that a 'sequential' test should be made for shopping proposals. These should focus on town centres, and only if shopping needs cannot be satisfied there should a search then begin on edge-of-town sites, and only if this fails to identify suitable sites, should out-of-town sites be sought. If an out-of-town site is necessary they should be assessed by four criteria: impact on vitality and viability of existing town centres; accessibility by a choice of transport modes; likely effect on overall travel patterns and car use; and the impact on development plan strategy. Elsewhere, the guidance argued against regional shopping centres and factory shops, but underlined the crucial role of village shops and urged planners to be sympathetic to them and to take their importance to the community into account when considering applications for change of use to residential. Most critics however agreed that the advice was bolting the stable door long after the horse had bolted with many out-of-town shopping centres either having been built or waiting to be built, and that it failed to reconcile the irreconcilable, the public demand for out-of-town car based shopping (Denison, 1996).

The advice was influenced by a House of Commons Environment Committee (1994a) report on shopping centres which advocated that out-of-town shopping centres should only be allowed if they could pass the 'sequential' test that no alternative sites were available working from the town centre outwards. The Government broadly accepted the 'sequential' test in its response (Cm 2767, 1995) to the report and promised to revise PPG6 to emphasise the need for a plan-led approach. The Environment Committee (1997b) welcomed the revised advice in PPG6 but expressed concern that the revised guidance was failing to give town centres the protection they needed and also concern about large scale out of town leisure developments.

The advice in the 1994 version of *PPG13 Transport* was followed up in two ways. First, the Department of the Environment (1995f) issued a 'Guide to Better Practice' which contained detailed advice on how to reduce dependence on the car. Two years later they issued (Environment, 1997c) a supplement which implemented proposals in the Transport Green Paper *Transport-the Way Forward* (see Chapter Six) to bring trunk road planning into the regional planning guidance system and how all those involved in regional planning guidance should relate this to the integration of land use and transport planning.

PPG24 Planning and Noise (Environment, 1994d): replaced Circular 10/73 and advised local planning authorities to separate potentially noisy developments from noise sensitive areas.

A useful index of the PPGs has been provided by the Council for the Protection of Rural England (1995c). The entry for the re-use and adaptation of farm buildings for example cites PPGs 2,7,12,15,17,21 and 22 as sources of advice with the paragraph or Annex that the information can be found.

Circular 18/94 Gypsy Sites Policy and Unauthorised Camping (Environment, 1994e): dealt with the impact of the Criminal Justice and Public Order Act 1994 which included powers to prevent trespass by 'New Age' Travellers. The Circular advised that in certain cases a policy of toleration may be applied to gypsies camping on unauthorised sites, notably where their removal would lead to a greater nuisance being created on another unauthorised site. Nonetheless, gypsies should live on authorised gypsy sites.

Circular 9/95 General Development Order Consolidation (Environment, 1995g): Provided advice on the consolidation of the 16 changes made to the GDO since 1988 when it was last consolidated and a number of new changes. These included the change that development requiring an Environmental Assessment will no longer benefit from permitted development rights.

Circular 10/95 Planning Controls over Demolition (Environment, 1995h): provided guidance on the changes brought about by the 1990 and 1991 Acts and by the 1995 'demolition order' which brought total demolition of certain buildings within development control. Partial demolition is still allowed. In effect development control powers were introduced chiefly to demolition of houses *outside* conservation areas and to gates, fences and walls *within* conservation areas.

Circular 11/95 The Use of Conditions in Planning Permissions (Environment, 1995i): maintained the structure of its predecessor (1/85) but revised it in several important aspects, notably with regard to a number of court judgements, for example on agricultural occupancy conditions. It also incorporated guidance introduced by PPGs since 1985, notably with regard to transport, retail development, contaminated land, noise and affordable housing. Additional advice on design and landscape, holiday occupancy and nature conservation and endangered species was also provided. The six tests for conditions remained the same, namely: necessary; relevant to planning; relevant to the development; enforceable; precise; and reasonable in all other respects.

Circular 4/96 Local Government Change and the Planning System (Environment, 1996h): advised local planning authorities on how the changes in local government boundaries which came into force on 1 April 1996 would affect the planning process. In particular it advised on how to maintain the structure plan process in areas where the Secretary of State has indicated that strategic planning needs to be carried out over an area wider than that of individual authorities by preparing a joint structure plan.

Circular 13/96 Planning and Affordable Housing (Environment, 1996i): provided guidance on where it would be inappropriate for local planning authorities to seek such housing when it stated: '*in practice the policy should only be applied to...settlements with a population of 3,000 or fewer to developments of 25 or more dwellings, or to any residential site of 1 or more hectares irrespective of the number of dwellings and elsewhere to developments of 40 or more dwellings or residential sites of 1.5 hectares or more*' (p.3). In terms of rural sites the Circular advised that local plans should make it clear that such sites should be released as an exception to normal plan policies and that they could count as additional to general housing demand. Concern has been expressed about the abuse of the agricultural and forestry exceptions and so the Circular advised that local planning authorities should set an upper size limit in terms of floorspace or number of bedrooms and that such houses should be located within or adjoining existing villages.

Circular 1/97 Planning Obligations (Environment, Department of, 1997d): reworked the five tests invoked by its predecessor, Circular 16/91. In the current Circular, planning obligations should only be sought where they meet the following tests: necessary; relevant to planning; directly related to the proposed development; fairly and reasonably related in scale and kind to the proposed development; and reasonable in all other respects. Most commentators agreed (Healey, 1997 and Johnston, 1997b) that the new Circular gave obligations a more positive role to emerge as a valuable tool in regulating development and achieving planning policy objectives.

The Department of the Environment also published: advice on how businesses should best seek planning permission (1995j); policy guidelines for the coast (1995k) which failed to do much more than collect together existing policies (Huggett, 1996); a good practice guide on evaluating environmental information for planning projects (1994f); and an index of planning guidance (1995l). In addition regional guidance was issued for the North West, Yorkshire and Humberside and the West Midlands so all of England is now covered by regional planning guidance.

The Department of the Environment (1995n) also issued practical advice on how planners could encourage the rural economy. This included the need for local authorities to:

- assess the needs and opportunities for diversification in their areas;
- co-operate in compiling registers of rural buildings with unimplemented planning permission for economic re-use;
- inform unsuccessful applicants for planning permission what forms of development are acceptable under the development plan; and
- recognise the economic potential of projects which add value to local produce.

The Countryside Commission (1996g) published a consultation paper which called upon planners to seek more local character and distinctiveness from planning applications and to be less reactive by setting out five objectives for countryside planning: to retain a beautiful and diverse countryside; to minimise the loss of countryside to urban development; to conserve and enhance local distinctiveness; to meet the needs of people living in the countryside; and to keep the countryside as a place to live and work in and to visit and enjoy. In particular the planning system should replace satisfying demand for development by a presumption that the system should only meet need which could be tested by a set of criteria based on environmental, economic and social objectives.

In Wales the Welsh Office (1996) issued a document which consolidated existing planning policy guidance used in England and introduced some significant changes, notably, the sequential test for shopping developments and for the first time provided advice on Green Belts in Wales. This advised that Green Belts should only be considered for the more heavily populated areas of Wales, but that within them, development should be strictly controlled. The document was criticised for its brevity, but this criticism was to some extent countered by a series of Technical Advice Notes issued in 1996 on for example: Housing Land Availability Studies; Nature Conservation; Development Involving Agricultural Land; and a Planning Guidance Wales document on Unitary Development Plans.

The Appeals System

The *Town and Country Planning (Costs of Inquiries etc.) Act 1995* regularised the practice of charging local planning authorities for employing planning inspectors on development plan inquiries, which had existed since 1976 when the first local plan inquiry was held, but for which no legal basis had existed till this Act was passed.

Circular 15/96 Planning Appeal Procedures (Environment, 1996j): consolidated all departmental advice on inquiries and other appeal procedures in a single updated document. The Circular aimed at speeding up the system, reducing delays and expense and making the system more accessible and user friendly by promoting existing best practice. A consultation paper issued by the Department of the Environment in January 1997 went further down this route when it suggested that the choice of appeal method should be made by the Secretary of State. The paper also suggested ways in which the system could be speeded up, mainly by cutting down the number of responses that could be made, and the time allowed for responses.

In July 1996 the Department of the Environment decided that the Planning Inspectorate should continue as a Next Steps Agency within the Department and should not be privatised or contracted out.

Plan Making

The *Town and Country Planning (Development Plan)(Amendment) Regulations 1997* (S.I. 1997 No. 531) updated the 1991 regulations. The main change was to require local planning authorities to have regard to any future national waste strategy as trailed in Cm 3040 the 1995 White Paper on 'Making Waste Work'.

Proposals to speed up the plan making process were made in four consultation papers from the Department of the Environment in 1995, 1996 and 1997. The responses to these papers made it clear that Section 54A had made the process more protracted as landowners and developers were now more inclined to object and pursue issues at the inquiry. The 1997 paper proposed a number of short term measures based on best practice and a number of longer term measures which would need legislation. These included: the imposition of time limits; the replacement of the right for objections to be heard for a right for them to be considered; and more radically (Milne, 1997b), making the Inspector's recommendations binding on the local planning authority.

Local Government Review

The review of local government was completed in March 1996. The new authorities took power on 1 April 1998. The review created a complex system based on 46 unitary authorities and over 30 two-tier authorities, mainly the traditional rural counties, except for Herefordshire which regained its independence as a unitary authority. In counties like Devon which had two new unitary authorities and a new National Park Authority, it was expected that all the authorities would voluntarily work together to continue the structure planning process carried out by the former two-tier county.

Many commentators were very critical of the review and called it a shambles leading to an unsustainable mess which will soon have to be reviewed again (Delafons, 1995). Meanwhile in Wales a much more coherent and logical system of unitary authorities came into operation on 1 April 1996 based on 22 unitary authorities which replaced eight counties and 37 districts. Most of Wales was covered by 11 rural authorities many of which covered a very large area, notably Powys, while 11 small urban authorities covered a small area between Newport and Swansea.

Exhortation

The Labour Party (1995) proposed a phased introduction of regional planning. The first phase would be a regional chamber of nominated elected councillors. The second phase would be directly elected regional assemblies, but only after unitary authorities had been set up. The Association of Metropolitan Authorities (1996b) called for a more holistic and proactive view of planning with a third party right of appeal against planning permissions contrary to the development plan. More prosaically the Local Government Management Board (1995b) published a guide to sustainable settlement planning prepared by the University of the West of England (Steeley, 1995) which according to Fyson (1995) should become the most thumbed aide-memoire on a planner's desk, notably the quick summary checklist of over 200 bullet-points.

More romantically Fairlie (1996b) has written a book about a more sustainable style of home building for the longer term. This is based on the concept of *'Low Impact Development'* a social contract whereby people are given the opportunity to live in the country in return for providing environmental benefits. *'Low Impact Development'* is defined as one that through its low negative environmental impact, either enhances or does not significantly diminish environmental quality.

His text is divided into two. The first, is an assessment of the planning system and in particular its failure to provide living space for people who genuinely want to live *from* the land. The second, is a set of case studies and visions of how such people could easily be accommodated into the planning system as an experiment towards a more sustainable society. The book is strongly influenced by his inability to recreate in England, the self-built wooden shack and smallholding he created in France for nine years.

The degree to which these ideas will be espoused remain however problematic. It would be all too easy for critics to dismiss them as the posturings of an alternative lifestyle, and indeed the main problem with these ideas as with all genuinely sustainable lifestyles are that they are so far removed from our consumption based culture. Nonetheless, Fairlie is not proposing an overnight revolution but a chance for a minority group to be able to demonstrate realistic alternatives, while we buy time. It is to be hoped that they are given every chance.

In the shorter term a number of reports exhorted better design standards. For example, the Council for the Protection of Rural England (1995d) deplored the low design standards of much new housing in the countryside and called upon planners to reinstate local building styles. These views were endorsed by the Countryside Commission (1996h and 1997a) in a multi-media pack aimed at reintroducing local character into villages by better designs. Finally, Rudlin and Falk (1995) discussed ways in which houses and their location need to be more sustainable alongside the Department of the Environment's (1996k) discussion of *Sustainable Settlements and Shelter* produced for the 1996 UN Habitat II Conference in Istanbul.

Implementation (excluding Listed Buildings)

Agencies

The Planning Inspectorate Executive Agency (1995a and 1996) continued to meet its performance targets, albeit with some adjustments as the workload varied between different types of work. In their corporate plan for 1995-99, the Agency (1995b) warned that further increases in workload, notably in the development plan and environmental case work under the 1990 Environmental Protection Act could lead to performance targets not being met. Similar warnings were issued by the Advisory Panel on Standards for the Planning Inspectorate (1995 and 1996) while still acknowledging that the inspectorate achieves unusually high standards of service.

Research by Prime Research and the University of the West of England for the Department of the Environment (1995m) found that the public regarded planning to be a necessary and valued aspect of public policy, which provides value for money and leads to improved quality in development. Planning also received a good press in the Oxford Brookes analysis of responses to the *Quality in Town and Country* discussion document issued by the DOE in 1994. However, in six detailed areas of planning which included: mixed use, transport and density, views were far more varied (Environment, 1995n).

The first systematic study of elected members in the planning process was carried out by Roger Zetter (1997) of Oxford Brookes University for the RTPI. This revealed widespread concern following much publicised cases of abuse of the planning system in North Cornwall, Warwick and Ceredigon. Councillors were found to be particularly worried by lobbying, declaring an interest, and whether to vote as they saw fit or to represent dominant public opinion. The study argued that the three crucial factors governing planning decision making - probity, transparency and impartiality - must be seen to be beyond reproach. This should be achieved by codes of conduct and protocols which should replace some of the more lax forms of decision making found to operate in some of the smaller district councils.

Another survey of councillors by the Audit Commission (1997) revealed a wide disparity in decision making structures with some councils having a complex committee system with for example, some planning committees dealing in detail with every planning application, while others delegated as many as 80 per cent of decisions to officers. The Commission recommended that councillors should play a more representational role. At the same time research into the effect of the 1985 Access to Information Act by Steel (1995) found that planning meetings attracted the biggest and most regular attendance by the public, notably when controversial local issues rather than strategic issues were being discussed.

A longer term view of planning is provided by Cherry's (1997) valedictory text on *Town Planning in Britain since 1900* which provides a straightforward descriptive account of planning in the 20th century with no bias or theoretical concepts. It is good old fashioned story telling based around people and organisations. Nine of the book's ten chapters cover each main period beginning with a review of the *zeitgeist* of each period, followed by the planning changes in that period. Most of this material is however already well recorded, albeit not so elegantly.

The book's main contribution however rests with the final chapter which considers the place of planning now and in the future given the new post-collectivist consensus which implies a withering and sustained diminution of the 20th century notion of the planning ideal, in which the idea that man could shape the world around him is now seen as a 'fatal deceit' according to Hayek. If this is the case then Cherry concludes that Whitehall will tolerate planning as a harmless low-key activity which may be marginally beneficial as a useful balancing ploy among contending aspirants in environmental protection. Thus planning has been neutered and its visionary role has gone to be replaced by work which will be relatively low key, short term project-based and incrementalist and managerialist in style, rather than long term and strategic.

The result is thus a surprise free, highly regulatory, over-systematised set of arrangements, lacking entrepreneurship, novelty and creativity. It is also a system which upholds the 'public interest' which is in fact a bundle of 'self-interests'. Cherry thus argues that one of the critical questions for planning concerns the 'public interest'-its definition and who defines it. Other crucial questions are the lack of intellectual inquiry, imagination and inherent energy about what planning is trying to do. If these issues aren't faced then Cherry worries that the bread and butter work of planning: plan making and development control will be increasingly called into question as a costly, highly regulative system supported by an over-manned, self-protective bureaucracy.

A shorter time horizon is provided by Brindley, Rydin, and Stoker (1996) in *Remaking Planning; The Politics of Urban Change* which in its second edition keeps all the original material unchanged but adds a postscript chapter. This argues that the 1980s now look

more like part of a long term trend to postmodernity than a distinct Thatcher decade. There does therefore seem to be something distinctively different about current times in four areas:
1) A new spatial economy based on flexibility is using space more actively and threatening national and local states with outward disinvestment or promising inward investment;
2) A new politics is based on governance, networks, actors and partnerships;
3) Social fragmentation is replacing broad unified narratives with a polyphony of many stories told by many story tellers. Mediating local social relations is thus more difficult; and
4) The new environmental agenda emphasises uncertainty and risk and thus the dominant utilitarian calculus of classical planning is at risk. In contrast one response is to seek strategic planning options which call for traditional planning skills based on the public interest and synthesising and integrating information.

Two new types of planning have however emerged in the 1990s. First, in buoyant market areas *responsive planning* which combines aspects of trend and regulative planning. Second, in marginal to derelict market areas *partnership planning* combines aspects of leverage and popular planning. In contrast, private management and public sector investment planning have dropped out of the agenda.

For the future, planning as managing land use change will need to be more pluralistic and listen to others outside established pluralist groups. Planning will also need to understand process and local actors more than collecting data, will need to produce alternative scenarios, and will need to learn how to incorporate environmental issues in other ways than grand sustainable plans.

In a book of edited essays Tewdwr-Jones (1996) attempts to assess the way in which planning has changed in the 1990s. This period has seen considerable changes in the plan preparation process, notably: i) self-approval of structure plans and the introduction of district wide local plans; ii) the introduction of so-called plan-led development control under Section 54 A of the 1991 Planning and Compensation Act; iii) the increased scope and importance of PPGs; iv) the approval of Regional Planning Guidance Notes in most regions; v) proposals to reform local government; and vi) the increased importance of European and environmental issues.

The period also marks the transition from Mrs Thatcher to John Major, and in the opening chapter, Tewdwr-Jones notes that this has witnessed a radical overhaul of the framework within which planning operates, but not of policies. John Major has also made a mark in one other way, namely the Citizen's Charter, and its performance indicator approach to assessing public policy.

The book is written by a mixture of planners, academics, councillors and solicitors, and as such provides an interesting cross section. However, every contributor presents a rather bland and text-bookish account, and though this is disappointing, it may perhaps indicate a considerable consensus about planning's current strengths and weaknesses. The book is reasonably comprehensive in its coverage, but only considers the countryside in passing. In conclusion, this is a good book, especially for those seeking an *aide-memoire* of the last five or so years, not least for some of the insights given by the non-academic authors concerning the actuality of planning practice.

In another edited book, Greed (1996) examines *The role of town planning in the development process*. This is quite useful in parts, but almost entirely city oriented with a major focus on Bristol. The book is divided into six parts: 1) The developmental and professional context, which is interesting if standard material; 2) the legislative and procedural framework which is good on planning gain; 3) resolving conflict which is a very good section, notably on appeals and negotiation; 4) the process in practice which contains some good case studies; 5) people, policies and power which begins to go off the rails into 'other' issues; and 6) alternative perspectives which contains much speculation and exhortation, though much of it is quite sensible. Overall the book is rather chatty and lacks gravitas, copious references, theory or style.

A number of other books on town planning were also published in the period under review. Booth (1996) provided a comparison of different systems of controlling development in Europe, the USA and Hong Kong, Lucarelli (1995) provided an account of how some of Mumford's ideas have been applied, Franey and Company (1995) have provided a *Planning Factbook* and Trench and Oc (1995) have provided a review of some current issues in planning based.

Policy Guidance and Plan Making

Research for the DOE by Land Use Consultants (1995a) found that PPGs are a remarkably effective means of disseminating national policy priorities and had ensured a more consistent approach to development plan formulation. However, the research also revealed four areas where improvements could be made. First, by improving the consultation process. Second, by making them more user-friendly. Third, by providing more definitive guidance on sustainable development. Fourth, by integrating their advice better with other government policies. With regard to one specific PPG, draft PPG13, Land Use Consultants (1996) forecast that developers would desert green-field locations.

In the field of plan making steady progress was made towards achieving complete district wide local plan coverage, although about 10 per cent of authorities were completing their plans at least a year behind schedule. The Department of the Environment (1995o and 1995p) issued two reports: first, on the efficiency and effectiveness of local plan inquiries; and second, on community involvement in the planning process which reported on recent projects in which the public have become genuinely involved in consensus building plan making exercises. In more detail the House Builders Federation (1995) argued that environmental capacity was being spuriously used to reduce housing allocations in plans, and that it should only be used as a factor when it was based on explicit quantifiable criteria.

Development Control

The Department of the Environment (1995q) published research by Land Use Consultants which showed that abuse of the agricultural dwelling loophole as demonstrated by Gilg and Kelly (1997) is still occurring, but that the controls on the siting and design of farm buildings introduced in 1993 were helping to reduce their landscape impacts. A further study by Oxford Brookes University for the

Department of the Environment (1995r) revealed that planners were implementing policies in PPG7 which called for a more relaxed approach to rural diversification on farms, both in development plans and development control decisions. The study was accompanied by a good practice guide (Environment, Department of, 1995s) which emphasised the need for local planning authorities to: 1) assess the needs and opportunities for diversification; 2) co-operate in compiling registers of rural buildings with unimplemented planning permission; 3) inform unsuccessful applicants what forms of development would be acceptable; and 4) recognise the economic potential of projects which add value to local produce.

The Department of the Environment (1995t and 1996l) continued to publish estimates of land use change in England which showed that around 50 per cent of development in the early 1990s was taking place on previously developed or so-called 'brownfield' land. On average, 5,850 net hectares of rural land was lost to urban use, while 8,400 net hectares of agricultural land was lost to urban and other uses in 1990 and 1991.

A more complex analysis of the supply of housing land via the planning system is provided by Bramley, Bartlett and Lambert (1995) which according to Hooper (1997) provides a set of workable models for predicting the effects of different strategies. For example, widespread land release, even in the green belt would lead to only marginal reductions in house prices. In addition, the models can quantify the trade-offs that may occur between density, price and other key housing variables.

More pragmatically, the Department of the Environment (1995u) issued the findings of a research report by Roger Tym and partners which revealed that Article 4 Direction Orders are widely used and believed to be effective weapons by local planning authorities, notably in residential conservation areas. However, the research did recommend that a new circular should be issued on how to make best use of Article 4 Orders. The Department of the Environment (1995v) also issued research by Arup Economics concerned with how the planning system could prevent derelict land.

Morgan and Nott (1995) published a second edition of their 1988 text book on development control, which although it is a useful reference, falls between the two stools of an academic analysis based on research and the dryness and accuracy of a legal tome. More legalistically, Grant (1996) updates a previous text on 'Permitted Development' to take into account changes to the General Development Order between 1988 and 1995. Campaigners who desire to stop and influence planning permission will find useful advice in the book by Speer and Dade (1994) while those who wish to appeal are provided with advice by Banks (1994). Such advice may well be needed since research by Land Use Consultants (1995) has shown that only about 25 per cent of appeals were successful, but that success rates varied quite widely between areas. The research focused specifically on Areas of Outstanding Natural Beauty and found that success rates varied from 0% to 36.6% in the AONBs and from 0% to 39.4% in the surrounding areas.

Three Issues: 4.4 Million Households/Environmental Assessment/ Minerals

4.4 Million Households: The Great Housing Debate

In March 1995 the Department of the Environment (1995w) issued a new set of household projections which forecast that the number of households in England would grow from 19.2 to 23.6 million between 1991 and 2016, an increase of 4.4 million and nearly one million more than the previous projections published in 1991. Local authorities were told that the projections were not an estimate of the number of additional houses to be provided in local authority plans but were just one of the factors to be taken into account when arriving at figures for housing provision to be included in regional guidance and development plans. Further technical information for this debate was provided by the Office of Population Censuses and Surveys (1995) subnational projections based on mid 1993 data.

The projections were based on assumptions relating to increased immigration, more divorces and declining household sizes based on trends between 1971 and 1991. For example, one person households were forecast to grow by 3.5 million from 5.1 to 8.6 million. Overall household size was forecast to decline from 2.67 people in 1981 and 2.47 in 1991 to a mere 2.17 in 2016. The projections were greeted with alarm and while some criticised them as being exaggerated many planners argued that even if they were 25% incorrect they still indicated a massive demand for new houses and thus the need for an urgent debate about where the new houses would be built. Other commentators, for example Holmans (1995), argued that the projections were an underestimate and that around 250,000 new houses a year would be needed by 2011, considerably more than the annual rate of 176,000 implied by the 4.4 million forecast. In contrast, Bramley and Watkins (1996) criticised previous projections for creating self-fulfilling prophecies because new houses built to meet predicted demands encouraged new households to form.

The main debate however began 18 months later with the publication of a DOE Green Paper *Household Growth: Where Shall we live?* in November (Cm 3471,1996). This pointed out that the projections implied an extra 169,400 hectares of urban land between 1991 and 2016 which would raise the percentage of urban land from 10.6 to 11.9%.

In contrast to the panic reaction by some commentators and pressure groups, some more rational thinkers pointed out that the total projected increase of 169,400 hectares in urban land, implied a loss of 6,700 hectares rural land a year, which was much below the 15,000 hectares of land lost in the 1960s and much less than the 25,000 hectares of loss per year in the 1930s. In addition, the 4.4 million houses total implied an annual rate of house building of 176,000 houses per year, much less than the 250,000 to 300,000 houses per year built in the 1950s and 1960s.

Turning to the percentage of land that could be built on existing urban or 'brownfield' land the Green Paper then proceeded to ask for views on '*the value of an aspirational target of 60% or whether we could do even better*' (p.40). The initial reaction to the discussion paper by environmentalists was that the 60% figure was too low while property professionals said it was unrealistic, because the current rate was only 50% (see above). Indeed some commentators suggested that the 'brownfield' rate could fall to as low as 30% as surplus land in built up areas was used up. For example, Bibby and Shepherd (1996) in a technical report for the Department of the Environment demonstrated that there were enormous variations across the country in demand for new households and the supply of brownfield land. South west England would fare badly because it had a high demand but few 'brownfield' sites, while the North west would fare better because it had lower demand and plenty of surplus urban land.

In terms of special protection areas, for example, National Parks and Green Belts, the paper floated the idea that when local authorities prepared their development plans they could be encouraged to check that policies of restraint are not being maintained for reasons that are now out of date. Also in preparing local plans they might be willing to ask themselves whether all areas of protection from development and all existing levels of protection are still appropriate?

Responses to the discussion paper were varied. For example, Nick Raynsford, the Labour Party spokesman for housing in the March 1997 issue of *Town and Country Planning* recognised that the 4.4 million estimate was about correct, and that the planning system in the past had failed to modify household formation. In addition, the rate of 175,000 new houses implied by the projections was below the average numbers of houses built in the 1950s, 1960s and 1970s. Indeed, the low rate of house building in the early 1990s implied that more than 4.4 new houses might be needed. Raynsford thus argued that the 60 % 'brownfield' rate should not be seen as a national target or to justify town cramming as in the disastrous tower block estates of the 1960s. Instead the question should be one of creating good quality environments, not of reacting against all development as undesirable.

The Town and Country Planning Association based their response on the principles set out in their book on *Planning for a Sustainable Environment*, their report entitled *The People: Where will they go?* (Town and Country Planning Association, 1996) and the assumption that most people wanted to live in suburban or rural areas and remain dependent on the car. Accordingly they advocated a strategy based on concentrating growth where existing settlements could be aggregated, or linked by transport corridors. Other solutions could focus on the creation of New Towns, focusing growth into key villages and allowing decentralisation to villages where this would sustain local services under threat. The RTPI and the Local Government Association broadly agreed with Raynsford and the TCPA that the 'brownfield' target rate was not a sensible measure to focus on, especially given the century long trend to decentralise homes and jobs, which would take several decades to reverse, even if we wanted to and had the powers that might reverse the trend.

In contrast, the Water Companies Association warned that any more development in the south east would place severe pressure on water resources and could prove disastrous. The Council for the Protection of Rural England (1994 and 1996b) in a series of reports also argued that the projections and their antecedents would place unbearable strains on environmental capacity in many south east counties.

Related commentaries were also provided by a number of studies and reports. For example, Bramley and Watkins (1996) from a study of 162 districts found that environmental capacity was used by about two-thirds of them in determining housing targets, but that this hadn't prevented an increase of housing supply in both social and market housing. They also found that household numbers seem to have been affected by the supply and price of housing, although they also argued that house prices would only fall by a moderate amount even if a large amount of extra land were allocated for housing. The potential for the planning system to influence household growth therefore seems to be limited.

Other commentaries examined the degree to which more attractive or higher density cities could stem the flow of people to the countryside. For example, Jenks, Burton and Williams (1996) discussed the benefits and dangers of higher density urban living in sustainable compact cities, while the UK Round Table on Sustainable Development (1997) demonstrated that 75% of all new housing could be built within existing urban areas. The National Housing and Town Planning Council (1997) took a different approach when they advocated the regeneration of tower blocks to house one of the main factors in the projections, the growing numbers of single people.

The debate had been preceded in February 1996 by a House of Commons Environment Committee (1996d) report on 'Housing Need' and the Government's response in May 1996 (Cm 3259, 1996) which focused on social and affordable housing. Both documents welcomed the estimates of housing need and household projections and stressed the importance of further research into how household formation is affected by economic and social factors, as well as the impact of government policy. Both documents stressed the need for more behavioural input to the projections notably with regard to the ratio between owner-occupation and letting. Both documents considered the role of the planning system and the need for better co-ordination between the top down approach of Regional Planning Guidance and the bottom up approach of local development plans. Finally, both documents agreed in favour of increasing the % of brownfield development above the 50% level recently achieved.

Environmental Assessment Procedures

In June 1995, developments that did not need planning permission under the General Development Order, were brought under planning control if they needed an Environmental Assessment. Circular 13/95 (Environment, 1995x) which gave advice on the change anticipated that only a small number of developments would be involved, for example: field drainage works; coastal land reclamation; and local authority road improvements in designated areas like National Parks, or near Sites of Special Scientific Interest.

The 1985 Environmental Assessment Directive was amended by the Council of Ministers in March 1997 after a long period of discussion. Member states were given till March 1999 to implement the Directive. The amended Directive contained a number of significant changes. First, the amended Directive demands that all projects needing an Assessment should pass through a 'development consent' process. Although most projects pass through the UK planning system, there are 17 other statutory systems through which a project might pass, notably afforestation and salmon farming. It was expected that these would need to be brought under planning control. Second, a developer will have to outline the main alternatives and the main reasons for selecting the chosen project. Third, the amended Directive increased the number of projects subject to compulsory Assessment, for example, groundwater abstraction or artificial groundwater recharge schemes, work for the transfer of water between river basins, dams for holding back or storing water, installations for the intensive rearing of poultry or pigs, and quarries and open-cast mining. Fourth, a number of projects which could be subject to Assessment were also added, for example, deforestation for conversion to another land use and permanent caravan or camp sites.

The European Community also approved a draft directive on Strategic Environmental Assessment, in December 1996, which would extend the principles of Environmental Assessment into the policies, plans and programmes of public agencies. This would reduce the problem with current procedures which take place too late in the planning process. According to Thompson (1997) a weakness of the proposals was that the European Community would not be subject to the Directive, even though some of its programmes, notably the Structural Funds, can have significant environmental effects. Given the long gestation of Directives it is not however likely that the proposals will effect the UK till well into the next century. Nonetheless, in the meantime, a survey of local planning authorities by Oxford Brookes University revealed that about three-quarters had either completed or were in the process of completing 'Environmental Appraisals of Development Plans' as advised by the DOE in 1993.

A guide on how to carry out an environmental appraisal based on experience in Bedfordshire was published by the Royal Society for the Protection of Birds (1996a). It contained 9 main procedures:1) decide who will do the appraisal; 2) choose a comprehensive list of environmental factors; 3) decide on the approach, based on the type of document being appraised; 4) identify and assess strategic options; 5) carry out the appraisal; 6) revise the plan or programme in light of the appraisal findings; 7) appraise the revised plan or programme; 8) present the results of the appraisal; and 9) set up monitoring procedures to feedback into future reviews (Earthy and Dodd, 1996).

A number of publications were issued concerned with undertaking Environmental Assessment. For example, the Department of the Environment (1995y and 1994g) published a good practice guide on the preparation of environmental statements and a research report on existing good practice which was used to produce the good practice guide. The Environment Agency (1996c) also produced a scoping handbook. Morris and Therivel (1995) edited a useful and detailed collection of chapters based on the Oxford Brookes University MSC course on Environmental Assessment. As Environmental Assessment becomes an ever more used technique this excellent volume should find a place on every practising planner's shelf, and should allow them to carry out an EA. If planners don't feel confident enough to carry out an EA themselves this book will allow them to brief consultants effectively.

The Department of the Environment (1996m) published research by Oxford Brookes University which showed that the quality of Assessments had improved. Nonetheless, only about two-thirds of Assessments were deemed to be satisfactory. Similar figures were produced in a research project focused on wildlife issues for the Royal Society for the Protection of Birds (1996b). This research examined 38 environmental statements between 1988 and 1994 and found that only half had used survey and analysis techniques appropriately and that 57% had failed to clearly predict ecological effects. Nonetheless, some good statements had been produced and the report made 30 recommendations to bring all future statements to the standard of the best produced (Tyldesley, 1996).

More theoretically, Lichfield (1996) the acknowledged expert on cost-benefit style analyses of planning decisions has now assembled all his accumulated expertise in an attempt to develop the technique into a generic method of community impact evaluation. This method is based on a sophisticated expansion of the Geddesian planning process in which 16 boxes identify the tasks to be undertaken at each stage of the process from initial proposal to final decision. The basic idea is to provide all those involved in planing with a common measuring rod with which to evaluate development proposals.

The book thus represents a throwback to the period of rational planning models developed in the 1960s and 1970s and envisages at first sight an idealistic, logical world with time to spare, perfect information, and apolitical planners and politicians. Although such an approach is at odds with post-modern ideas it is to be hoped that Lichfield's book is studied carefully and his methods are further evaluated in order to see if they can provide a broad set of procedures applicable at all levels of the planning process in perhaps one last attempt to rescue planning from the defeatism of post-modernism on the one hand and mechanistic zoning on the other. If planning is about seeking a balance between all the factors then Lichfield has provided a potentially powerful tool by which to find the best balance.

Minerals

The Environment Act 1995 contained provisions forcing local authorities to modernise planning permissions given for mineral working between July 1948 and February 1982. Any new conditions which would restrict working rights or prejudice the economic viability or asset value of a site could lead to a claim for compensation. Advice on how to modernise old planning permissions was contained in Mineral Planning Guidance Note 14 published by HMSO in 1995. The definition of exactly what criteria should be used to assess whether compensation would be payable or not was however left vague by both the Act and the MPG. Instead the MPG argued that in the absence of case law, the words: economic viability and asset value have their common or everyday meaning. In the future all mineral sites will be subject to review and updating every 15 years.

Elsewhere, general advice was updated in 1995 by MPG1 which took account of legislative changes since the original MPG1 was issued in 1988, notably the need for development to be sustainable. New advice on peat provision was provided by MPG13 issued in 1995 which argued that strict controls should be placed on existing peat diggings and that new sites should only be allowed where they had already been damaged by human activities and were of limited or no conservation value. For the longer term the MPG set a target of 40% for non-peat materials to be used by the total horticultural market by 2005. Nonetheless, the MPG suggested that another 1,000 hectares of land might be subject to new peat extraction in the next ten to twenty years. Finally, advice on 'The reclamation of mineral workings' was issued in MPG7 in 1996.

Elsewhere, a survey for the DOE revealed that some 18,800 hectares of mineral workings or spoil tips had been reclaimed between 1988 and 1994, of which almost 90 per cent was due to planning conditions being attached to permissions. The Council for the Protection of Rural England (1996c) published a guide informing campaigners how to object to or modify applications for mineral planning permission which they deemed to be necessary since Owens and Cowell (1996) in a study for the Council had found that the aggregates industry via privileged access to the planning process was able to place the need for aggregates above the needs of the environment.

Scotland

Consolidation of Scottish Planning Legislation

Scottish planning legislation was consolidated into four Acts which came into force in May 1997. The four Acts were: The Town and Country Planning (Scotland) Act, 1997; the Planning (Listed Buildings and Conservation Areas)(Scotland) Act 1997; the Planning (Hazardous Substances)(Scotland) Act 1997; and the Planning (Consequential Provisions)(Scotland) Act 1997. No major changes to legislation were introduced although new procedural rules for planning inquiries were made.

Review of the Planning System

In 1994 the Scottish Office issued a consultation paper inviting views on how the planning system in Scotland might be improved. In 1995 the Scottish Office issued a reply to the responses entitled: 'Review of the Town and Country Planning System in Scotland: The Way Ahead'. This focused on improving performance targets for both plan making and development control via best practice methods aimed at streamlining the system. The consultation process, reported in a related publication 'Summary of Responses to the Consultation' revealed however, not only a desire for more efficient procedures but also quality decision making involving local communities which should be addressed by further reforms, albeit in the context of the present system, because the majority of respondents believed the basic framework of planning in Scotland to be satisfactory.

In September 1996 a number of measures designed to improve performance and encourage higher standards in the delivery of planning services were announced. These included: new targets for processing applications; a new planning award scheme; and new research into the issue of quality assurance.

Changes to the System: Local Government Areas and Plan Making

Scottish Local Government was radically reorganised on 1 April 1996 when the two tier system of 9 regions and 56 districts was replaced by 32 unitary authorities. However a two-tier system of structure and local plans continued. In most areas unitary authorities prepare both types of plan, but in the central belt 3 structure plans will be prepared by groups of unitary authorities, for the Lothians, Strathclyde and Ayrshire, with individual unitary authorities preparing a local plan. In all 16 Structure Plans were expected to be prepared. Some doubts were cast on this potentially confusing situation at the time (Hayton, 1996) but it will be a number of years before the new system can be evaluated, especially since the Scottish Office issued a revised NPG3 (see below) in November 1996 which offered unitary authorities greater flexibility in determining structure plan land allocations for housing. The Scottish Office also issued a Code of Practice for Local Plan Inquiries in 1996 which updated and doubled the advice given in the previous 1985 Code. The Scottish Wildlife and Countryside Link (1994) published a report which recommended that the new authorities should set up an Environment and Sustainable Development committee and work with other authorities to produce regional countryside resource strategies.

Actual and Proposed Changes to the System: Development Control

Three consultation papers were issued by the Scottish Office in 1996. First, a Green Paper was issued on possible changes to listed building procedures and on related heritage legislation which would improve a sense of partnership between central and local government and between the public and private sectors. Second, a consultation paper on planning inquiries proposed to improve the efficiency and effectiveness of the inquiry process while maintaining the essential requirements for openness, fairness and impartiality. Third, a consultation paper on the Use Classes Order included a proposal to allow the limited use of a house for bed and breakfast, which might help farm tourism by removing the need to obtain planning permission for occasional use, defined as no more than 90 nights a year. The arrangements for the review of old mineral permissions, outlined above came into force in Scotland, on 1 January 1997. Finally, in March 1997 the Office proposed a number of relaxations for planning controls over telecommunications antennas.

Advice: NPGs, PANs and Circulars

NPG3: Land for Housing (1996): Gave planning authorities greater freedom to decide the allocation of land in their structure plans compared to the original 1993 version. Structure plans should nonetheless meet full housing demand where practicable and reasonable, but demand assessments should not be the sole consideration and greater account should now be taken of other factors including economic, infrastructure, environmental and amenity considerations so that some small towns and villages can be protected from inappropriate development.

NPG7: Planning and Flooding (1995): Emphasised that susceptibility to flooding is a material consideration.

NPG8: Retailing (1996): Stressed the need for a sequential test based on searching outwards from city centres. However, new out-of-centre shopping areas may be acceptable in certain cases. Although the NPG took 10 years to be issued, only a year after it had been published a revised draft was issued in March 1997 which proposed extra powers to local authorities in restricting the nature of goods sold and the size of units in a bid to give added emphasis to town centres.

NPG10: Planning and Waste Management (1996): In 1996 the Scottish Environment Protection Agency (See Chapter Two) assumed responsibility for environmental regulation in Scotland with a duty to prepare a National Waste Strategy by 1998 to replace local authority waste disposal plans. This NPG provides advice on how Structure and Local Plans should be co-ordinated with this strategy.

Draft NPGs were also issued on Skiing Developments in 1996, Sport, Physical Recreation and Open Space in 1996, Transport and Planning in 1996, and on Coastal Planning in 1997.

PAN37: Structure Planning was revised in 1996 and stressed the need for structure plans to focus on key issues and to provide concise and clear policy statements. However, Hayton (1997) argued that the Note was flawed in that it failed to provide on how to produce joint structure plans or how to involve the lay community more fully.

PAN38: Structure Plans: Housing Land Requirements was revised in 1996 and confirmed the advice in NPG3 (see above) that housing demand as measured by household projections should be tempered by other factors, for example, infrastructure sustainability at the local level.

PAN47: Community Councils and Planning gave detailed advice about how unitary authorities should inform community councils about planning applications, how they should be expected to respond (see Circular 4/96 below) and how they could be involved in local plan making.

PAN49: Local Plans which replaced PANs 30, 32 and 34 in 1996 introduced advice as to how to incorporate sustainability principles into local plans. It also urged planning authorities to produce local plans in 3 years and in due course to alter plans in 2 years, but it remained ambiguous about the scope and range of issues to be included.

PAN50: Environmental Effects of Mineral Workings attempted to introduce sustainable principles to the control of mineral workings.

Circular 6/95: (see Chapter Five).

Circular 4/96: Reminded unitary authorities of their new duty to consult with community councils over all planning applications.

Circular 12/1996: Superseded the advice given in Circular 22/84 concerning planning agreements. Agreements should: serve a planning purpose; be related to the proposed development; be related in scale and kind to the proposed development; and be reasonable.

Circular 34/96: Gave detailed advice on how to review old mineral planning permissions given between 1948 and 1982 (see above) with dates by which each phase should be completed. It was expected that imposing short time safeguards would be completed by 1 January 2000, with more detailed reviews to follow, followed in turn by periodical reviews every 15 years.

Circular 4/97: Updated and consolidated advice in Circular 29/88 over those planning applications that should be notified to the Scottish Office, for example, certain applications in nature conservation or 'National Scenic Areas'.

Circular 13/97: Provided advice on changes to the procedures for planning inquiries and hearings to make the process more efficient and equitable.

4 Extensive Land Uses

Agricultural Production

The Wider Context

Starting at the global level, the 1992 GATT agreement was formally adopted by the Council of European Agricultural Ministers in December 1994 in terms of implementing the detailed arrangements on import quotas and export refunds in 1995. Talks on further freeing up agricultural markets were due to begin in 1999 under the auspices of the World Trade Organisation, the successor to GATT, with a view to reaching agreement early in the millennium.

In the USA the 1996-97 Farm Bill was forecast to have a dramatic effect on not only US farmers but also on world trade in agricultural commodities. The Bill effectively abolished short term set-aside, and offered farmers decreasing cash payments decoupled from production and capped at a maximum of $50,000 per farmer. The Bill thus threatened to force thousands of small cereal farmers out of business as the new freedom to farm offered by the Bill allowed big farmers to expand production, thus leading to a dramatic fall in prices. Things could have been worse if 15 million hectares (equal to the whole EU wheat area) of the 22.5 million hectares set-aside under the long term Conservation Reserve Programme (CRP) had not been retained till 2002. The general downward trend of US prices to world prices and the subsequent weakening of world prices was forecast to place even greater pressure on the CAP to further reform itself and reduce financial support to market prices.

Within Europe, talks concerning the enlargement of the European Union continued. The House of Lords Select Committee on the European Communities (1996) produced a report on the enlargement of the European Union and CAP reform which broadly endorsed the European Commission's view that the CAP would have to be fundamentally reformed and farm prices reduced to world market levels if enlargement was to be a realistic option. More details on this report and other reports on European agricultural policy are to be found in Chapter One.

Within the existing European Union it was acknowledged that each of the three forces outlined above underlined the need to continue the reforms made in 1992. The UK remained one of the prime movers behind the pressure for continued reform. Accordingly, the Ministry of Agriculture, Fisheries and Food (1995a) published a plan which proposed a gradual reduction of production-linked support and end-prices, and a corresponding removal of quotas, set-aside and similar controls. In contrast more should be spent on developing a sustainable agriculture to be promoted via a mix of economic instruments, advice and regulation backed up by a substantial level of public funding.

A longer term perspective was provided by Gardner (1996) an author with many years of experience with the CAP. Although Gardner provides a useful overview of the CAP, the book is repetitive, lacks any conceptual or classificatory rigour, is based too heavily on reports and conference papers rather than in-depth research, and speculates too much on the possible impacts of the 1992 reforms. Nonetheless, the book contains some powerful anecdotes about the CAP and some powerful visual aids to understanding the Byzantine complexity of European decision making.

Fennell (1997) has according to Grant (1998) produced a thorough and well informed systematic historical study of the development of the CAP, based very heavily on European Commission documents. It is thus a rather indigestible and top-down approach. Nonetheless, it is an authoritative and comprehensive study which will be most useful as a reference book for years to come.

Turning to the future, the NFU (1995) published a discussion document which advocated lowering commodity support prices in order to move closer to world prices, but supporting farmers with aid decoupled from production support. Similar proposals to decouple support were made by ADAS in 1995. The Country Landowners Association (1994) not only proposed decoupling, but also advocated cutting back on supply controls and other regulations as agriculture returned to a free market. The market would however be cushioned by whole farm area payments which would over a 10 to 15 year period come to include a growing emphasis on environmental management payments as farmers drew up contracts to provide environmental services. Finally, the European Environmental Advisory Council (1997) recommended that environmental objectives should be put explicitly at the heart of a reformed CAP.

Agencies

One-third of ADAS's work was privatised via a management buy out in 1997. At the same time the statutory, mainly agri-environment, work of ADAS was continued as from April 1997 by a new agency, the Farming and Rural Conservation Agency (FRCA) under the auspices of MAFF. The new agency was expected to provide advice to all levels of government on a wide range of issues, varying from structure plans, major planning applications, how best to deliver specific schemes like ESAs, and the development of national agricultural policies (Cm 3604, 1997).

For the future a number of calls were made, notably by the Labour Party in 1996, for a food standards agency, independent of MAFF in order to avoid the clash of interests between consumers and producers that had manifested itself during the BSE crisis. In January 1997 the Government responded by announcing that it would set up a food safety council but this fell foul of the election.

Legislation

The *Agricultural Tenancies Act 1996* gave landlords and tenants greater freedom to decide the terms and length of their own tenancy agreements in the hope that it would encourage more letting of farmland. In particular the Act enabled tenancies to cover a wider range of rural enterprises, in which farming is the main but not the exclusive business. The main farm organisations welcomed the Act as a substantial change in the way that farmland is let, but some experts in the field of property law doubted if the Act would recreate the farming ladder of former years (Bullen, 1995c) and that the flexibility in the system would deter landlords from offering land because of the need to negotiate very detailed watertight agreements (Gibbard, 1995). Another boost to letting land was however provided in 1995 when land let under a farm business tenancy was exempted from inheritance tax to bring it into line with owner-occupied land which had been given similar relief in 1992.

The *Transfer of Crofting Estates (Scotland) Act 1997* gave power to the Scottish Office to sell its 109,000 hectare crofting estate to the crofters. Disposals can only be made to a body approved by the Secretary of State for Scotland, expected to be community-based crofter's trusts. These trusts will have to be approved by the Crofters Commission after consultation which will have regard to the general interests of the crofting community and the views of crofters in the district. The sale should save the Scottish Office money, since the estate only yielded £150,000 a year in rents compared to running costs of £370,000. The Act made no difference to the 16,200 crofters who rent their crofts privately.

Agricultural Policy Measures

Useful information on policy measures can be garnered from the Annual Reports and Expenditure Plans of MAFF (Cm 2803, 3204 and 3604) the Intervention Board Executive Agency, and ADAS. Indeed the House of Commons Agriculture Committee (1995a) has congratulated MAFF on the production of a 'highly informative document'. Nonetheless, the Committee did ask that the reports should in future provide a fuller discussion of MAFF's overall strategies and the economic and social effects of agri-environment payments, notably, ESAs. In their reply (House of Commons Agriculture Committee, 1995a) the Government agreed to provide further evaluations as they became available, for example the five year evaluations of ESAs.

There were no major changes to price support in ECUs, but the dramatic fall and then rise of sterling between 1994 and 1997 led at first to increased support prices and payments and good farm incomes. As sterling recovered however in 1996 and 1997 the Green Pound was revalued leading to large falls in support for UK farmers and much reduced incomes. In more detail, the Green Pound rate in October 1994 was 0.78 ECU to the Pound, this rose to a high of 0.86

in November 1995 before falling to 0.74 in March 1997. Generally, the price cuts agreed in 1992 were implemented with a claimed saving of 82 million ECU in 1998. Further cuts were proposed, notably in arable area aid, to save a further 1.4 billion ECU in order to fund the beef emergency package caused by the BSE crisis, but the EU could not agree on these proposals.

The broad rate of set-aside varied from 15% for the 1994 harvest, to 12% in 1995, to 10% in 1996 and to 5% in 1997 with detailed variations depending on the type of set-aside. In line with sterling's volatility, set-aside rates varied; for example, from £341 per hectare in 1995 to £248 in 1996. In 1995 the EU approved a Regulation which allowed land entered into agri-environment schemes to count for set-aside purposes as long as it had been eligible for arable area payments, that is, it must not have been under permanent pasture, permanent crops or trees or used for non-agricultural purposes as of 31 December 1991. In 1996 the requirement to rotate set-aside around the farm was removed. Production of cereals fell from 22.1 million tonnes in 1992, the last year before compulsory set-aside, to 19.9 million tonnes in 1994, but then rose to 21.7 million tonnes in 1995, illustrating the insensitivity of supply controls based on area quotas.

As a result of the fall and rise of sterling, arable area aid payments were volatile. For example, in June 1995 arable area aid in England was £269 per hectare, £76 higher than 1994, but by September 1996 they had fallen to £247 per hectare. Around 3,600,000 hectares of land was under the Arable Area Payments scheme during the period. The IACS scheme continued throughout the period.

Intervention stocks were reduced during the period according to reports by the Intervention Board Executive Agency (1994, 1995 and 1996) but expenditure remained high at £2,700 million in 1995-96. Nonetheless, the reductions provided hope that intervention stocks might have been coming under control. However, much of the falls in beef and butter stocks could well have been due to the BSE crisis and the emergency measures introduced to remove cattle from the market.

The Department of the Environment designated 68 Nitrate Vulnerable Zones in March 1996 under the 1991 European Nitrate Directive. The Zones had been vetted by an independent review panel in 1995 which rejected some zones and altered the boundaries of others, but confirmed most of the proposed zones. It was expected that the regulations controlling land use in the zones would be introduced in 1999. Most of the largest zones were in East Anglia, the Midlands and Yorkshire. It was estimated that the annual cost to farmers would be £3 million with a one-off cost of £10 million as they set out to reduce nitrate levels to the target rate of 50 mg/litre. In the existing 32 Nitrate Sensitive Areas land became eligible as set-aside as from 1995. Earlier in 1995 the 10 pilot NSAs had been incorporated into the existing voluntary 22 NSAs launched in 1994. Research for MAFF found that farm income in the pilot areas had increased by £55 per hectare. Elsewhere the scheme to promote farm waste management plans in certain selected river catchments was extended to a further seven areas in 1995.

Organic farming continued to grow in popularity spurred on by the introduction of an Organic Aid scheme in 1994 as part of the 1992 agri-environment regulation (Candy, 1994). However, uptake of the scheme was lower than expected and so in July 1996 MAFF launched a £150,000 organic information service to encourage farmers to find out more about organic farming. Around 900 responses were received in the nine months to April 1997, three times more than expected (McDougal, 1997b). Nonetheless, only 600 farmers were qualified as organic farmers in 1997, covering a mere 50,000 hectares. As a result about 70% of organic food was still being imported.

The Potato Marketing Scheme and its associated quota on potato plantings were wound up at the end of the 1996 crop year, on 30 June 1997. As a result the Potato Marketing Board was replaced by the British Potato Council which is funded by a levy on growers and first purchasers. In advance of the changes, many producers negotiated contracts to sell their potatoes so that at least 50% of sales were contracted (Abel, 1995).

Milk farmers who had already experienced the reform of the Milk Marketing Scheme faced uncertainty as the milk prices offered by the main purchasers fluctuated between them as the new system established itself. Quota continued as a double-edged weapon. On one side it offers a tradable asset, but on the other side it restricts expansion of production with the threat of levies for over production.

Beef farming was severely affected by a dramatic new development to the long running BSE (Bovine Spongiform Encepholy) crisis when MAFF announced in the spring of 1996 that CJD (Creutzfeldt-Jakob Disease) in humans could be caused by eating BSE infected beef. Exports of British beef were soon banned and a collapse in consumer confidence led to a series of related measures costing an expected £1 billion in the first year alone. First, a cull programme to kill up to a million cattle over 30 months old to prevent them entering the food chain at a cost of around £550 million in the first year. Second, £80 million in top up payments for prime beef cattle over 30 months old. Third, £80 million to abattoirs to dispose of male calves that used to be exported. Fourth, £80 million to buy and destroy 40,000 tonnes of beef already in surplus stores; fifth, £30 million in emergency aid for extra slaughtering; and sixth, £118 million in aid to rendering plants. At the end of 1996 the Budget announced further spending: £300 million for the over 30 month scheme; £245 million for intervention support buying; £95 million for slaughterers and renderers; and £12 million for research and development. In order to provide support for farmers who sold cattle younger than 30 months at below expected prices a series of beef marketing schemes at a cost of around £60 million were also introduced during 1996.

The BSE crisis which in the 1980s and 1990s had been characterised by a series of measures which were too little and too late, was thus transformed into a crisis in which the measures were probably far too much too soon. It did however reflect the classical British response to policy making, incremental and reactive. The full story of the BSE episode when it is fully revealed will tell us much about British rural policy making. BSE cost £1,370 million in 1996-97 in MAFF expenditure alone.

The Hill Livestock Compensatory Allowance Scheme was modified in a number of ways, notably in terms of the support offered. In 1995 the rates remained the same as in 1994, but after inflation the real rate of support continued to fall to about half its value in 1980. The rates were also held for 1996. However, in November 1996 the rates for 1997 were altered to take account of the BSE crisis. As a result rates for cattle in severely disadvantaged areas rose by £50 to £97.50 and in the disadvantaged areas by £46 to £69.75. Rates for sheep in the uplands were kept the same at between £5.75 and £2.65 per ewe, depending on the type of ewe.

Farmers and Farming

Farm incomes rose by nearly 5 per cent in 1994 to reach their highest level since 1984. Farm incomes then rose very significantly, by around a third, in 1995 to reach their highest level in two decades helped by the continued weakness of sterling. Within the sectors, cereals recorded incomes three times the early 1990s average, followed by general cropping with twice the level at the beginning of the decade. The only sector to show a decrease was pigs and poultry. However incomes began to fall in 1996 with an average drop of nearly 10%. Every sector however recorded higher incomes than the early 1990s. Nonetheless, farm employment continued to fall, with a 3.7% drop between 1991 and 1996.

Perspectives

An official perspective on the future of British agriculture was provided by a report from the Technology Foresight Panel (1995) within the Cabinet Office's Office of Science and Technology. They predicted that fewer farms would produce higher quality food from less land. Spatially, farming will become more diversified with some areas specialising in industrial 'non-food' crops, some areas specialising in recreation and conservation areas, and some areas concentrating on competitive food production. In a similar approach, but at a world scale Pretty (1995) has identified how three types of agriculture have emerged. Two of his models: industrialised; and low-input traditional farming, are relevant to Britain. He argues that if agriculture is to be sustainable at the world scale, that policies to encourage low-input methods should replace those policies that assume the globalisation of the industrial model, because this could not be globally sustainable. Lee (1996) has argued that Pretty's ideas provide an inspirational vision for policy makers.

In the same vein Baldock *et al* (1996) have produced a report for the CPRE and the WWF which set out the broad principles and detailed actions needed to provide a more sustainable agriculture. This began by developing a model which placed agriculture at the fulcrum of three sets of demands: demands from society for natural resources; demands which agriculture places on the natural environment; and demands which society places upon agriculture. In addition there are the constraints imposed by global trends. Given this context there are a number of practical steps which can be taken now which include: lower inputs; reducing farm wastes; improving water quality; altering stocking patterns; and recreating landscapes and habitats. In the longer term there are five options for the future: a) reduce CAP subsidies; b) encourage intermediate production systems; c) encourage low input/organic farming; d) encourage the revival of mixed farming; and e) encourage more regional self-reliance.

The sustainable use of soil was examined in a thorough text book style report by the Royal Commission on Environmental Pollution (1997). The overall message of the report is that the soil is a vital resource which needs detailed and sophisticated integrated management, on a par with the air and water. The creation of the Environmental Agency has created an ideal opportunity to remedy past neglect and to give soil its rightful place in the protection of the environment. Another publication which demonstrated the necessity for soil management and ways in which this essential natural asset could be maintained was provided by Taylor *et al* (1996) in the context of Scotland.

More generally, the future of Scottish agriculture was considered by a major report from the House of Commons Scottish Affairs Committee (1996). Most of the recommendations were accepted by the Government (Cm 3548, 1997) as either reflecting existing policy or policy commitments. However, the report was not convinced that agriculture should necessarily be the recipient of decoupled payments justified on environmental and social grounds and suggested in common with the Scottish Rugby Union the adoption of New Zealand procedures, notably the New Zealand Resource Management Act. The Government response in this area was less enthusiastic and argued that agriculture is still seen to be the key plank in managing rural landscapes and economies as reflected in both UK and EU policies.

A more polemical critique of farming trends was provided by Harvey (1996) in his graphically entitled book '*The Killing of the Countryside*'. He demonstrated how the British public have not only lost the countryside as a precious national asset, but have also suffered the indignity of paying for the destruction via agricultural subsidies. According to Potter (1997) the diagnosis is powerful, but the prescription is less compelling based as it is on the premise that removing all farm support will lead to consumer pressure for an organic system, rather than a further ratchet of agri-technology as farms become ever larger and more capitalised in order to compete in a global market freed from subsidies and trade controls.

In the academic arena Lowe, Marsden and Whatmore (1994) have edited a collection of essays on '*Regulating Agriculture*' which although it includes chapters from around the world also contains a good deal of material of value to British rural policy makers. In particular, the opening chapter by the editors brilliantly assesses the move away from state controls towards market-based forms of regulation. This process has however been beset by contradictions and apparently perverse outcomes, for example, deregulation has often been accompanied by re-regulation. For the future, the book picks out two key trends. First, the rise of international trading agreements and free trade areas, and second, the rise of transnational food corporations who can pick and choose where and how they produce food all around the world.

More prosaically the Rural Development Commission (1996a) published research by the Centre for Agricultural Strategy which forecast that the 1992 CAP reforms would lead to the loss of 5,400 jobs within farming, 6,600 jobs in upstream jobs and 13,900 downstream jobs.

A number of House of Commons reports were also published into detailed aspects of agriculture. Milk quotas as a tradable commodity were examined by the House of Commons Agriculture Committee (1995b) as a response to the big increase in quota prices in 1994-95. The Government in their reply (House of Commons Agriculture Committee, 1995b) welcomed the Committee's conclusions that the increase was the result of market forces rather than any manipulation of the market by speculators or quota agents. The report was followed by a more wide-ranging report by the House of Commons Agriculture Committee (1996) into the UK dairy industry and the CAP dairy regime. The Committee were convinced that a radical reform to a free market system was both inevitable and desirable. In particular, a reformed industry should be less reliant on market support mechanisms, more responsive to consumer demand, more competitive in world terms and less constrained by artificial supply controls, notably, milk quotas. The Government, in their reply (House of Commons Agriculture Committee, 1996) noted that when the radical reforms were achieved there would be a need for financial support to shield the dairy industry from expected financial upheaval.

The House of Commons Agriculture Committee (1995c) also produced a report on horticulture which has a very adverse balance of payments with imports of £3,120 million only very partially offset by exports of £302 million in 1993. The Committee nonetheless gave the sector an optimistic prognosis and restricted itself to a series of detailed recommendations aimed at helping UK producers most of which were accepted by the Government in their response (House of Commons Agriculture Committee, 1995c).

Finally, the National Audit Office (1995) examined river pollution from farms and found that although the number of all pollution incidents reported to the National Rivers Authority had grown since 1985, the proportion relating to farms had dropped from 17% to 9% in 1993. This was an overall drop of 12% from 3,050 to 2,680 between 1985 and 1993 for farm pollution incidents. This partial success story was attributed to preventative measures designed to prevent farm pollution which have included targeted campaigns and farm visits, and also getting farmers to produce Farm Waste Management Plans using grants under the Farm and Conservation Grant Scheme.

Land Management

Changes to Schemes and Grants and Evaluation of Impacts

Existing powers to make land management grants to farmers and others (Countryside Commission, 1995k) were extended by the Environment Act 1995 which gave appropriate Ministers, with the consent of the Treasury, the power to make regulations for the making of grants to persons who may undertake work that will be conducive to conservation, namely:

'*(a) the conservation or enhancement of the natural beauty or amenity of the countryside (including its flora and fauna and geological and physiographical features) or of any features of archaeological interest there; or*
(b) the promotion of the enjoyment of the countryside by the public.'
(Section 98)

Environmentally Sensitive Areas continued to be the Government's flagship scheme and the 5,000th agreement was reached in November 1994. The 1997 MAFF report (Cm 3604, 1997) reported that ESAs were largely successful. In 1995 payments in 11 of the ESAs were increased by an average 7.5% at a cost £1 million a year. In 1996 another six ESAs were given payment increases of 15% at an annual cost of £1.5 million a year. Also in 1996 five ESAs were given new types of payment and were increased in area. By mid 1996 the English ESAs covered over a million hectares and involved 7,700 farmers. However at the end of 1996 the budget for ESAs was reduced from £69 to £59 million because of a lower than expected take-up. Nonetheless in February 1997 payments in 11 ESAs were increased by an average 8%.

The Countryside Stewardship scheme was transferred from the Countryside Commission who had pioneered it to MAFF on 1 April 1996. At the same time expenditure was raised by £5 million in 1996/7 and 1997/8 on top of the £11.25 million available to cover the annual cost of the 5,300 agreements entered into during the first five years of the scheme since 1991. Under the relaunched scheme ten year management agreements were offered for a wider range of work which would maintain, enhance or restore landscape and scenic beauty, wildlife habitat and historical and archaeological features. In February 1997 a pilot scheme was launched in which arable farmers were paid to manage arable land according to specific requirements, including leaving cereal stubble over winter and providing crops for wildlife cover and food (McDougal, 1997c).

The success of the Countryside Commission's pioneering scheme was revealed by a survey (McDougal, 1996b) by Wye College and the University of Reading which estimated that 3000 jobs had been created, that 90% of farmers had maintained or increased their incomes, and that traditional country skills had been given a boost. Small farmers capitalising on farm tourism had benefited most, while large arable farmers had suffered losses due to reduced inputs and extensification. Another study by the Centre for Environment and Land Tenure Studies (1996) found that Stewardship payments had increased land values by around 4% on good land, by 6% on poor land and by 8% on average land.

The Moorland Regeneration Scheme was launched in March 1995, as the sixth and last of the 1992 agri-environment reforms to the CAP. In its first year it offered £25 per ewe removed from hill flocks as long as at least 20 hectares of land containing at least 25% heather were entered into the scheme and 10 ewes were removed. The scheme was open to all farms in the Less Favoured Areas unless they were also in an ESA. Around £10 million was allocated to the scheme in its first year. However, only 15 participants were attracted in the first year. Accordingly, rates were raised to £30 per ewe in 1996 and similar grants to those available in the Countryside Stewardship scheme were introduced.

As reported above the EU agreed in 1995 that land under agri-environmental schemes could be counted as set-aside land. However, a survey for the MAFF by ADAS found that 33% of farmers knew about only one of the schemes that encouraged long term set-aside, namely, the Farm Woodland Premium Scheme (Allan, 1995).

The Farm and Conservation Grant Scheme which had been introduced in 1989 closed in early 1995 in England, although applications for one-off small conservation grants under the non-plan side of the scheme were continued in connection with some of the post 1992 agri-environment schemes till February 1996. Committed expenditure of £23 million was incurred in 1996-97. Grants for installing or improving facilities for handling farm waste under the Scheme were however kept for those farmers in the Nitrate Vulnerable Zones which were designated in 1996 (see above). Funds from the Scheme were therefore transferred to the Zones.

In Scotland, the ceilings on the amount of money that could be claimed in ESAs rose from between £10,000 and £30,000 to £20,000 and £82,000 in 1996. Outside the ESAs a new Scottish Countryside Premium Scheme was introduced in 1997. The scheme was an integrated system which offered eighteen management and 12 capital works options along the lines of the English Countryside Stewardship Scheme. Examples of the options included: £125 for managing grassland for birds; and £1 a tree for amenity tree planting. Expenditure ceilings of £30,000 per IACs business were set over five years. The scheme was warmly received by Scottish Natural Heritage who earlier (1994) had advocated financial support for active environmental care by farmers.

In Wales, Tir Cymen, the Countryside Council for Wales farm stewardship pilot scheme was modified in 1996 to make it more directly relevant and appealing to owners of intensively managed grass and arable farms. A survey by ADAS in 1996 had shown that the scheme had created more than 200, mainly casual, jobs and that farms were collecting on average £1547 in whole farm payments and £2510 in annual management payments. In terms of overall incomes it was estimated that farm incomes would have fallen by £864 without the scheme, and the net effect of the scheme had been to increase income by £1616. It was also estimated that there had been a substantial increase in environmental work in the pilot areas, notably in stone walling and woodland work.

Longer Term Evaluations of the Schemes and Recommendations for Change

The Ministry of Agriculture (1995b) issued a consultation document which sought views on the development of environmental land management schemes based to some extent on a preliminary report on the Countryside Stewardship Scheme by Land Use Consultants which was attached to the document. This recommended that: 1) Countryside Stewardship should continue to be based on the key principles established during the pilot phase; 2) Countryside Stewardship and ESAs should be developed to become MAFF's two core schemes; 3) schemes could perhaps be targeted towards arable areas; 4) further scheme mergers should be considered once the agri-environment schemes have been operating long enough to be evaluated; and 5) in the meantime a first stop information service should be offered for farmers interested in environmental schemes.

In their response the Countryside Commission criticised Seas for being too backward looking and seeking to fossilise landscapes. Instead, the Commission called for a more positive and forward looking approach which actively promoted certain habitats and landscapes. Alexander (1995) and Swales (1995) welcomed the proposals but Swales did so only as a first step towards much greater funding and a move towards attaching environmental conditions to funding under the much larger CAP budget.

A House of Commons Agriculture Select Committee (1997) report on ESAs and other agri-environment schemes made a number of general and detailed recommendations. In general the Committee recommended that: 1) geographically all the agri-environment schemes should be combined into a single national framework; 2) more research is needed into the impact of the schemes, notably the economic impact both on farms and on CAP budgets via reduced farm output; 3) MAFF should increase expenditure on agri-environment schemes but this increase should not match the decline in production subsidies; 4) the introduction of payments to farmers in accordance with environmental achievements is not a realistic proposition; but 5) agri-environmental payments should only be paid to farmers exceeding basic standards set out in a proposed Code of Good Agricultural Practice for Wildlife and Landscape; and 6) urgent consideration should be given to introduce cross-compliance requirements for arable area payments, in spite of a report described by Bullen (1995d) which revealed little support for linking farm support payments to environmental objectives.

The Committee also made a number of detailed recommendations: 1) the countryside access scheme should be closed because it had been a dismal failure; 2) MAFF should consider whether the objectives of the Nitrate Sensitive Areas Scheme could be achieved by other means, in the meantime tree planting incentives could be introduced; 3) increased payments could be made to farmers converting to organic production but subject to conditions relating to enhancing environmental features; 4) there should be incentives to encourage farmers to move up from tier 1 schemes in ESAs; 5) given the shift in planned resources away from ESAs to the Countryside Stewardship Scheme no further ESAs should be designated and in the longer term the two could be amalgamated; 6) there should be an opt-out option after 10 years for the habitat scheme to improve uptake; and 7) there should be increased payments under the moorland scheme if it can be rescued.

The Council for the Protection of Rural England (1995e) examined how the 1992 agri-environment regulation had been implemented in five other European countries as well as the UK. They found a considerable diversity of practice and that a more decentralised system based on local environmental needs would be feasible in the UK. Accordingly, the Council argued that there was a strong case to devolving the design and management of schemes to the local level. Local responses to agri-environment policies have also been studied by Winter, *et al* (1996) in their study of the conservation advice offered by ADAS and FWAG staff and by Winter, Watkins and Cox (1996) in their study of game management in England.

Hedgerows in England and Wales

The Environment Act 1995 gave power for Ministers to make regulations to make provisions to protect *'important hedgerows'* (Section 97). The Act stated that '*important hedgerows*' shall be determined in accordance with prescribed criteria and that the regulations may also make provision in relation to other hedgerows for the purpose of facilitating the protection of 'important hedgerows'. The regulations to implement the Act were issued in March 1997 and came into force in June 1997.

Under the regulations (Rusling, 1997) farmers should notify the local planning authority if they intend to remove a hedgerow. The local planning authority have 42 days to respond, after which the hedgerow can be removed. Within 42 days the local planning authority can either give approval for the removal, or issue a 'hedgerow retention notice' if the hedgerow is 'important'. The definition of important can only be given to hedgerow over 30 years old. After that a number of criteria may be used, for example, diversity of hedgerow species, adjacent to a footpath, provides a diverse habitat, or marking ancient features or ownership or parish boundaries. Planning and environmental organisations castigated the new regulations as being weak and unenforceable (Bartram, 1997 and Ward, 1997) and in the light of this it was perhaps not surprising that the NFU gave the regulations a cautious welcome (Hirst, 1997).

Forestry

Changes to the much criticised consultation arrangements for planting, felling and restocking were made in August 1996 (Forestry Commission, 1996). For new planting the Forestry Commission was asked to consult with local authorities for applications for 10 hectares or more, but only to consult over felling licences if the trees are covered by a Tree Preservation Order. Proposals for planting five or more hectares in National Parks were made subject to consultation by the National Park Authorities and in other designated areas, the appropriate national countryside agencies have to be consulted. The public register of planting proposals introduced in 1993 was extended to cover felling proposals.

The Forestry Commission Research Division became an Agency of the Forestry Commission in April 1997. Its first Corporate Plan for 1997-2002 set out two purposes for the Agency. First, a capability to conduct research, development, surveys and related services relevant to the forest industry, and second, to provide authoritative advice to support the development and implementation of the Government's forestry policy. The Agency will thus continue the work carried out by around 300 staff which has become increasingly focused on matching forestry to its environment, as demonstrated by the 'Report on Forest Research for 1996'. This reported on research into ecological site classification and research into the suitability of lowland sites for forestry at the 10 kilometre square level.

Three annual reports by the Forestry Commission (1994, 1995 and 1996) provided information about the changing nature of forestry planning. For example, the grant-in-aid needed by the Commission fell from £89 million in 1993-94, to £76 million in 1994-95 and to £59 million in 19995-96. The period also marked the transition of Forest Enterprise to an Executive Agency as from 1 April 1996. The report for 1995-96 noted that the Forest Enterprise made a net contribution of £17.9 million and had a cash income over expenditure of £11.5 million, with timber sales of £95 million. The 1995-96 report however also noted disappointment at the level of private planting, notably the decline in broad-leaved planting and hinted at a move away from grants to targeted incentives at identified opportunities.

Scotland continued to dominate planting with over half the total planting and the different position north of the border is reflected in the annual reports of the Scottish Office (Cm 2814, 1995; Cm 3214, 1996; Cm 3604, 1997) which provide an alternative source of information on the work of the Forestry Commission.

The Countryside Commission (1994f) published the strategy for the National Forest. The Strategy was implemented by the new National Forest Company from April 1995. Its aim is to plant 30 million trees, at the rate of 6,000 per day, at a cost of over £50 million with eventually one-third of the 194 square miles being afforested. It was hoped that 70% of the target area of new planting of 13,200 hectares would be achieved by 2005. Planting grants in the forest are based on prospective participants making competitive bids using the Forestry Commission's woodland grant scheme as a basis. In the first two years the tendering scheme was oversubscribed. By July 1996 830 hectares had been planted representing 1,183,000 trees, with access being secured over 540 hectares.

Five more Community Forests were approved around Manchester, Liverpool, Nottingham, Middlesborough and Sheffield in 1994 and two more around Swindon and in Bedfordshire were approved in 1995. A study of people's perceptions of the Community Forests carried out by Burgess for the Countryside Commission (1995k) found that though people valued forests for their sense of enclosure this also lead to an erroneous fear that forests were dangerous places to be avoided except in groups. Accordingly, the study proposed ways in which forests could be made to appear less threatening by providing three types of wood: an open wood for the timorous; a middle wood for those needing some sort of assurance; and a wild wood for people confident about wildlife and nature. A practical guide to creating multi-purpose woods around towns was provided by the Forestry Authority (1995).

For the future, the Countryside Commission and the Forestry Commission issued a consultation paper in 1996 on meeting the Rural White Paper target of doubling the woodland cover in England from 7.5 to 15% by 2050. The paper launched a debate on: what purposes the new woodland should have; the types of woods needed; where they should be sited; and how the expansion might be achieved. English Nature had already contributed to the debate in a 1995 policy statement on sustainable forestry and woodland management which argued that creating new woodland could be beneficial if it followed a number of principles based on interdependent multiple use. Some of the methods involved in managing forests for biodiversity were discussed in a set of symposium papers (Ferris-Kaan, 1995) which included: making an inventory of the diversity of certain taxonomic groups; identifying appropriate management units; considering linkages between habitats; and promoting tree crop diversity.

The differential costs of private forestry were revealed by Mitchell *et al* (1995) in a survey for the Forestry Commission. This revealed that the two most influential factors affecting the cost were the type of planting and the size of operation, with substantial economies of scale. The lowest costs were incurred in Scotland with its large investment plantings. In contrast England was the most expensive area where forestry is found in small holdings and mainly on farms and traditional estates. Woods managed for amenity, conservation or wildlife incurred the greatest expenditure, because of their small size and more intensive management.

In Scotland, tree planting under the Farm Woodland Scheme surpassed expectations in the first three years of the scheme with 14,200 hectares being planted on 875 farms by the autumn of 1996. The Millennium Commission announced £6 million to fund the second phase of the Millennium Forest for Scotland with an additional 32 forestry sites and the Scottish Office (1997a) published a review of the way in which indicative forestry strategies have encouraged the plantation of forests in an environmentally sensitive manner. An earlier review of nine indicative forestry strategies (8 in Scotland) by the Royal Society for the Protection of Birds (1996c) welcomed the approach but argued that inadequate protection for nature conservation was being provided. Nonetheless, the Society believed that the strategies would become a key component in delivering Government forestry policy, especially if backed by a national forestry strategy. For the future Gill (1996c) in a review of two discussion papers (Callander, 1995 and Worrell and Callander, 1996) on Scottish forestry argued that the synergistic value of European forestry where social, environmental and economic benefits come together might be achieved if woodland cover in Scotland were raised to 25-30% and if more native species were planted since these demanded more labour intensive management.

5 Nature Conservation and Recreation

Nature Conservation

Legislation and Advice About its Implementation

The *Deer (Amendment) (Scotland) Act 1996* amended the powers of the Deer Commission to reduce deer numbers where they are damaging the natural heritage, agricultural production or woodland growth. The *Deer (Scotland) Act 1996* consolidated the Act and former legislation relating to deer. Internationally, the European Convention on the conservation of wildlife was amended by the addition of around 20 mammals (Cm 3244, 1996). Amendments to the floral species and fauna species protected by the Berne Convention on the conservation of wildlife were also made (Cm 3583, 1997) and amendments to the convention on wetlands of international importance came into force in 1994 (Cm 3053, 1996).

The Friends of the Earth launched a symbolic Wildlife Bill in 1994 which became a private members Bill in 1996. This Bill would have prevented damage to SSSIs from pressures ranging from quarrying and peat digging to overgrazing and simple neglect. However, although the Bill completed all its House of Commons stages including the addition of Government amendments it ran out of time in the House of Lords. Another failed Bill, the Wildlife Bill 1996 would have made further provision for the protection of SSSIs.

The Department of the Environment (1994c) issued *PPG9 Nature Conservation* which advised on the planning implications of the regulations which implemented the EU Habitats Directive which came into force on 30 October 1994. The PPG also gave advice about how to integrate nature with economic development outside nature conservation sites, and advised that where conflict is unavoidable the adverse effects on wildlife must be minimised. The

PPG is considered more fully in Chapter Three and the Habitats Directive more fully later in this chapter.

Scottish Natural Heritage issued a Circular (6/1995) on how local authorities should implement the 1979 Birds Directive and the 1992 Habitats Directive which reminded them of the need to refuse permission for developments which are likely to have a significant effect on the integrity of designated nature conservation sites. The RSPB criticised the Circular as inadequate and also the slow rate at which important bird conservation sites had been designated in Scotland compared to the UK as a whole.

Agencies

English Nature was given a vote of confidence in its ability to run its own affairs after the Department of the Environment had carried out its first Financial Management and Policy Review. As a result English Nature was given greater independence from the DOE and a more strategic relationship was set up between the two bodies.

English Nature also published a number of progress and annual reports. For example, in February 1997 it published its 5th Annual Report for 1995-96. This recorded that 81 new Sites of Special Scientific Interest had been designated in the year bringing the total to 3,874 and the area to 920,696 hectares. The report also highlighted how it is seeking to provide more positive management in the SSSIs by developing relationships with landowners and land managers and by developing the Wildlife Enhancement Scheme (which became an England-wide scheme in April 1996) and the Reserves Enhancement Scheme. English Nature aims to have positive management plans for the vast majority of SSSIs by the millennium.

The policy of moving away from compensatory management agreements to positive management was also implemented in other areas of English Nature's work, notably in general management agreements and the further development of the Species Recovery Programme with 50 Species Action Plans prepared in 1995-96. Nonetheless, negative controls via 19 Section 29 Orders under the 1981 Wildlife and Countryside Act had to be made compared to 17 in the previous year and damage was recorded to 121 SSSIs compared to 104 the year earlier. During 1995-96 English Nature was given grant-in-aid of £36 million and a further £2.4 million to run the Joint Nature Conservation Committee. £23 million was spent on staffing and operational costs, £8 million on management agreements and £2 million each on National Nature Reserves, Grants, and Conservation Support. English Nature also continued to draw up more Statements of Intent with a wide variety of organisations in order to achieve its aim of nature conservation through partnership and co-operation.

In Scotland, the Scottish Office published a revised statement of Natural Priorities for Scottish Natural Heritage in November 1996. This emphasised (Ross-Robertson, 1997) the importance of Scottish Natural Heritage maintaining a balanced, broad-based, multi-functional approach to its work. This will mean continuing to balance nature conservation with the conservation of landscape and amenity, namely, obligations to designated areas with concerns about the wider natural heritage, while promoting understanding and enjoyment of the countryside. This will still have to be done through the voluntary principle and by seeking to form partnerships with other organisations.

Areas

The House of Commons Committee of Public Accounts (1995) examined SSSIs. The Committee urged English Nature to complete the notification process, which might lead to between 180 and 370 more sites, as soon as practicable. The Committee were broadly happy with the monitoring of sites but was concerned about how damaged sites were analysed and thus argued for a more comprehensive and robust system. The Committee endorsed the efforts made by English Nature to improve relations with site owners and occupiers but argued that not enough was being done to monitor management agreements or to prosecute or apply for Nature Conservation Orders where appropriate. Finally, the Committee expressed concern about substantial regional variations in completing management plans. Evidence on SSSIs was submitted to the House of Commons Environment Committee (1994b).

In a later and more specific report on SSSIs, English Nature (1996) indicated that 89 of England's most important wetland sites were adversely affected by water abstraction and that nearly 50% of these could be at high risk. English Nature accordingly sought an agreed way forward with the Environment Agency and the water companies to rely less on aquifer and summer water abstraction from rivers and more on winter storage and efficient water use. The Friends of the Earth (1994b) also issued a report which used data collected from a variety of local and national organisations to show that more than 10% of SSSIs are under threat.

Progress continued to be made with the designation of Special Areas of Conservation (SACs) under the 1992 Habitats Directive which was transposed into British law by the Conservation (Natural Habitats) Regulation 1994. In March 1995 a list of 280 potential SACs in the UK was sent out for consultation. In June 1995, a first tranche of 136 sites was submitted to the European Commission. A further 75 sites were submitted in January 1996. However conservation groups were worried that some key sites were omitted. In response the Government announced that it was considering further additions to the list in addition to existing programmes aimed at submitting more sites as Special Protection Areas under the 1979 Birds Directive and Wetland Sites under the Ramsar Convention.

In a very rare move in the other direction, English Nature announced in September 1996 that Braunton Burrows in North Devon was no longer a National Nature Reserve. This was because they had been unable to renew the sub-lease on the land in terms that were acceptable to English Nature and the landowner. Despite the change English Nature sought to ensure that the conservation value of the dunes was maintained via its status as only one of three UNESCO Biosphere Reserves in England and its status as a potential SAC.

In another loss for nature conservation, the Lappel Bank case in the Thames estuary, the European Court of Justice ruled that Britain had acted illegally in allowing part of the land - due to be designated as a Special Protection Area under the 1979 Birds Directive - to be used for the storage of imported cars, when in 1993 it had concluded that economic considerations should prevail over the wildlife significance of the area and had thus excluded half of the area from the designation process. The RSPB welcomed the decision and argued that decisions on the 136 internationally important bird areas waiting to be designated as SPAs would now have to be based purely on their wildlife importance.

The importance of non-statutory sites of importance was stressed in a 1995 position statement from English Nature which offered help and encouragement from anybody or any organisation who wished to maintain or enhance the conservation value of the sites. Most notably in this area, the National Trust for Scotland acquired the 30,000 Mar Lodge estate in the Cairngorms in 1995 with the aid of a £10 million grant from the Heritage Lottery Fund. Scottish Natural Heritage in partnership with other bodies drew up a management agreement to ensure that the estate will be run in a sustainable and integrated way. Five year management plans will reduce moorland damage caused by red deer by reducing red deer numbers, which will help the natural regeneration of the native Caledonian Pine forest.

Biodiversity Challenge

The Department of the Environment (1995z) published two reports from the UK Biodiversity Steering Group set up by the Biodiversity UK Action Plan in 1994. The first volume proposed a series of targets and proposals as a basis for a co-ordinated and effective forward programme till 2010. The second volume contained three lists. First, a long list of 1250 species causing conservation concern, second, a middle list of 286 species and 24 habitats for which action plans were being produced between 1995 and 1997, and third, a short list of 116 species and 14 habitats for which action plans had already been produced. Action plans to conserve the 116 species and the 14 habitats would cost £3.8 million and £12.9 million respectively in 1997 and £2.4 million and £37.2 million by 2010.

The Government (Cm 3260, 1996) accepted the targets and proposals in May 1996 and began the process of finding voluntary and private sector organisations to 'champion' the 116 species and 14 endangered habitats in the immediate action plan. By the end of 1996 13 awards had been made under a new Species Action Grants Scheme totalling some £75,000 which paid half of the costs incurred by organisations embarking on an Action Plan to conserve species which included: several species of butterfly, the great crested newt and the brown hare. In February 1997, ICI became the first corporate 'champion' when it provided £55,000 to aid butterfly conservation. Barkham (1996) welcomed the initiative as a welcome demonstration of government commitment to conservation but nonetheless feared that these small positive actions could be as sand against the tide given global trends acting adversely against nature.

The Action Plan followed earlier reports by the Royal Society for the Protection of Birds (1995) which set out detailed targets for 600 species for 35 habitats in a comprehensive blueprint for action from the voluntary conservation sector and realistic ways in which Government policy could be influenced in the favourable post-Rio climate (Tompsett, 1995 and Toogood, 1995). Methods to monitor the success of biodiversity action plans were provided by a consortium (World Wide Fund for Nature, 1996) which proposed that two indices, one a common species and one a rare species should be related to form a 'green gauge' of success or failure.

Practical Guides and Manuals

A number of guides and manuals were published. They included advice from the Joint Nature Conservation Committee (1994) on how to produce and deliver a strategic framework for plant conservation including how to survey and monitor species, how to produce management and recovery plans and how to raise awareness and involvement amongst the public. The Department of the Environment (1996n) provided advice on how to cost the conservation of biodiversity through the maintenance, restoration and re-creation of key habitats. In more detail, the Department of the Environment (1996o) provided guidance on how to manage damaged land for nature conservation, for example, by setting aside some of the land for conservation while developing the remainder for profitable uses so that some of the income could be used to conserve part of the site.

English Nature (1995b) published guidance on how the difficulties caused by the fragmented nature of much of England's natural habitats could be ameliorated, initially, by keeping them in good heart, and then by looking at the possibilities of enhancing and enlarging them, for example, by linking up the remnants of heathland, creating new woodlands on set-aside land and managing field margins and roadside verges in ways to complement the dwindling stocks of old grassland. English Nature (1994) had earlier published an overview on environmental sustainability which provided information on the general principles involved and examples of good practice.

Studies of Nature Conservation Issues

The Royal Society for the Protection of Birds (1996d) in a report on the use of planning conditions involving nature conservation, claimed that their survey of 560 cases, revealed that only 12 per cent sought to protect nature conservation conditions and fewer than one per cent sought to provide genuine enhancement. Most conditions merely sought to mitigate the impact of development and were anyway rarely monitored. Another problem was the lack of ecological awareness amongst planners. The Royal Society for the Protection of Birds (1996e) in a report on water abstraction revealed that 100 rivers and 200 SSSIs were in danger from drying out as a result of demands from agricultural irrigation and the public supply.

The Joint Nature Conservation Committee (1997) published a report which demonstrated that pesticides are continuing to indirectly reduce the number of farmland birds by removing elements of the food chain. Dramatic declines in the past 25 years were noted for tree sparrows (89% down) turtle doves (77%) bullfinches (76% and song-thrushes (73%). Nonetheless, some surveys reported (The Times, 3rd October 1996) rises in some bird species which was attributed to set-aside, although these surveys were offset by others which showed an accelerating loss of birds in the 1990s (The Times, 18th April, 1996). The particular case of game birds and their management was studied in a research report by Winter *et al* (1996).

Commentaries, Polemics and Visions

Beginning with a vision, Bill Adams (1996) has been writing on conservation issues for over a decade, but this book on a vision for

nature conservation is his best and most useful contribution so far. His central thesis is that conservation needs to take a radical review of itself and should start to look forwards rather than backwards. In particular this means concentrating less on what we have inherited and focusing more on what we might have.

We should thus abandon the more static and ordered concept of ecosystems and move to a more chaotic concept of a dynamic mosaic of patches-within-patches in which equilibrium states are the exception to the rule. In summary, we need to change our scientifically based view of nature and to replace it with a socially constructed view in which conservation is not about trying to stop the human impact on nature but about negotiating and managing change.

In developing these themes Adams provides a beautifully written history of the conservation movement, its legislative progress, and its main techniques centred on species protection and area designation. The book flows logically to its final chapter which argues forcefully for a bolder view of conservation based on regional/national landscape plans in which creative conservation, including species reintroduction and massive reafforestation with native trees figure prominently. Adams concludes with four key principles. First, we need to maintain diversity, second, we should build room for nature into economic life, third, we should build connections between people and nature, and fourth, we need to allow nature to function and create conditions for it to do so. As Adams admits the challenge for putting these principles into practice will be considerable. Adams' contribution has been to show the conservation movement that it can move out of the straitjacket of area designation and that it can think the unthinkable of recreating wild woodlands across our austere bleak uplands. It is now time to begin the long task of convincing everyone involved that creative conservation is the way forward.

One step along the road to creative conservation could be made by Conservation, Amenity and Recreation Trusts which are the subject of a useful book by Dwyer and Hodge (1996). They describe the rapid growth of this neglected land management sector which they call- CARTS- or Conservation, Amenity and Recreation Trusts, and which are non-profit organisations devoted to the ownership and

long-term management of land for environmental or amenity benefit. The sector is now so important that its estimated annual turnover of £110 million is double that of English Nature and the Countryside Commission combined. In addition they own 2.7 per cent of the UK, employ 4500 people, and have nearly 4 million members. The sector includes the National Trust, which accounts for about half the sector, the Royal Society for the Protection of Birds, and the Wildlife Trusts. Dwyer and Hodge also examine a range of other trusts before discussing their utility.

Dwyer and Hodge conclude that CARTs have considerable potential beyond their main aim of benign land management and argue that they should continue to develop relationships both with each other and with the State in order to provide alternative systems of countryside management. In conclusion, this book draws attention to a neglected but increasingly important sector and provides a useful overview, albeit in a long winded populist style, but this is appropriate in discussing organisations which depend on popular appeal.

In a much more populist style Pennington (1996) queries the dominant view which attributes countryside destruction to 'market failure' and which seeks a solution via centralised planning operated by benevolent managers. In contrast, Pennington argues that it is in fact 'government failure' that is pervasive, and therefore seeks a return to full private property rights. He argues that 'government failure' is due to the self interest of those involved in perpetuating and extending their policies and a lack of sufficient information on which to base their actions. In contrast, freeing the private sector from planning controls and subsidies would reveal the true value of each piece of land and lead to its positive conservation backed up where necessary by covenants. Pennington's views are controversial and although there is a good deal of truth in his diagnosis, his prognosis is naive and only sketchily thought through.

In a more traditional account Green (1997) has produced a third edition of his book on '*Countryside Conservation*'. Like its predecessors, of 1981 and 1985, it is oriented towards ecological aspects of conservation, albeit with sections on planning and management policies as well some sections on concepts. It is divided into three parts: Principles; Policies and Practice. Each of these parts

contains some good material, notably the querying of some basic assumptions, e.g. that biodiversity is good. Overall there is much familiar material here and furthermore there is a lack of an overall unifying concept to keep the text focused rather than episodic. Nonetheless, the book remains a good overview and the concluding chapter, albeit brief, does contain three succinct themes for the 1990s, namely: targeting resources; emphasis on financial incentives; and supplementing top down policies with grass roots action. Finally, the book ends with a plea for a clearer statement of conservation objectives and for the current window of opportunity caused by the crisis of the CAP to make some sites sacrosanct, to restore and create habitats and to move from negative to positive thinking.

Finally, Rodgers (1996) has edited a collection of essays on nature conservation and countryside law which *inter alia* include essays on the law of habitat protection, environmental gain and set-aside, forestry law and the environment, access to the Scottish countryside, and nature conservation and access to the countryside. Overall, John Alder in his essay argues that because most common law evolved before conservation became an issue, that it is axiomatically often found to be wanting. Therefore, if the environment is to be protected by legislation, Alder argues that a well understood justification for conservation and preservation must be developed (Bough, 1996).

Recreation

Legislative Change

Many organisations expressed fears that Section 68 of the *Criminal Justice and Public Order Act 1994* would fetter recreational access to the countryside. Section 68 created a new offence of 'aggravated trespass', under which a person may commit an offence if they trespass on land with the intention of disrupting someone else's lawful activity. The Section had been brought in to prevent hunt saboteurs, new age travellers and people attending 'raves' or pop music festivals from disrupting rural land uses. However, it was feared (Mattingley, 1994) that landowners might put up signs which might say: '*Shooting in progress. Trespassers may be prosecuted*

under the Criminal Justice and Public Order Act' and that these signs may be left up for a long time thus acting as a deterrent. However, by 1997 there were few signs that the Act had had much impact on legitimate access. As a counterweight, the Countryside Access Scheme (see Chapter Four) one of the Agri-Environment schemes came into operation in 1994.

In Scotland, the Scottish Office issued NPPG11 on 'Sport, Physical Recreation and Open Space' in 1996 which reminded local councils to maintain the Right of Way network, to identify and prioritise gaps in the network as part of a wider access strategy for sport and recreation, and to consider enhancing the network by Orders, Agreements or designations on public land. In the urban fringe and woodland, consideration should be given to using recreational sites as a buffer to farmland, establish new woodlands for recreation, allowing land use changes from agriculture to sport or recreation. Damaging developments in designated landscape, cultural heritage or nature conservation areas should not however be allowed.

The Environmental Impact of Leisure Activities

The House of Commons Environment Committee (1995c) published a major report on how leisure activities affect the environment. Their main, and to some people very controversial, finding was that tourism and leisure do not pose a serious, immediate or intrinsic threat to the environment and that there is no need to impose any general restrictions on leisure activities in order to provide blanket protection or the rural environment. The second main point made by the Committee was that visitor numbers had not increased significantly.

The Committee did note however that leisure uses of the countryside could cause problems, but that these were cultural rather than physical. The Government in their reply (HC 761 (94-95)) accepted this point and called for a local response based on shared responsibility where cultural conflicts arose.

In more detail both documents endorsed the need to have the footpaths and rights of way network fully operational by the year 2000, and the need to pursue access agreements with vigour. Both

documents also endorsed the economic value of leisure and tourism to the countryside and the need for planning to be sympathetic to such developments as long as they were well designed and respected the local environment. Both documents also broadly agreed that leisure and tourism developments should be sustainable and be based on developing existing resources, and that noisy and other activities likely to lead to conflicts might best be situated in areas of derelict land or in the urban fringe rather than in the countryside. However, the Government was less in favour of Committee recommendations that new golf courses should be subject to greater environmental scrutiny.

In spite of the general consensus between the two documents and a welcoming paper by Harrison (1996) which applauded the approach of seeking the 'Best Available Place', some commentators (Pritchard, 1995 and Elson, 1995) were able to report that deep divisions could still be discerned. For example, Elson argued that readers comparing the Committee's report and a report by the Council for the Protection of Rural England on *Leisure Landscapes* might be excused for thinking the writers were on different planets.

Rural Tourism

The Countryside Commission (1995l and 1995m) in partnership with three other bodies published a major advisory report on Rural Tourism based on the experience gained in fifteen projects. Four principles for rural tourism were set out. First, to minimise the impact on the global environment. Second, to sustain the local environment by setting limits to visitors and by restoring and maintaining the resources on which tourism depends. Third, to sustain the host community by encouraging activities which retain visitor spending in local communities. Fourth, to sustain the visitor by recognising the visitor's right to enjoy the countryside and to enable access where possible and appropriate. Sustainable tourism was projected as an activity rather than a set of activities, but the report did contain a seven point checklist. This included: transport policy aims to reduce the use and impact of cars; marketing aims designed to influence the scale and type of tourism in response to environmental and local factors; and visitor welcome aimed at influencing visitor behaviour by reducing negative impacts and increasing environmental awareness.

Access Issues

The Countryside Commission (1994g) issued a booklet which explained to farmers the provisions of the Rights of Way Act 1990 relating to the ploughing of land and cropping of land crossed by rights of way or other unsurfaced highways. More specifically, the Commission (1995n) published a booklet aimed at farmers involved with the Countryside Stewardship Scheme and in particular with the Education Access option designed to encourage school visits. For the future, the Commission (1996i) published a discussion document which summarised the progress made and identified the barriers towards attaining the national target for rights of way. More specifically, the Commission (1995o) had already published a consultation paper on how to provide a more cost effective way of managing the National Trails which outlined three options.

English Nature issued a position statement in 1996 on recreation and nature conservation. This argued that new leisure activities should be sited close to existing public transport, cycling and walking routes and the places people live in order to reduce car journeys. With regard to its own input, English Nature will be guided by the need to sustain the country's natural heritage, and pledged that when it worked with those promoting and providing recreational opportunities that it would ensure that natural qualities are retained. In more detail, English Nature will provide scientific and practical information on the potential impacts of recreation on wildlife and natural features, use National Nature Reserves to demonstrate good practice, and encourage greater use of Environmental Assessment.

In Scotland, the Scottish Office launched a national Concordat on Access to Scotland's hills and mountains in 1996. This set out a number of principles for the use of hill land and pledged co-operation between visitors and local communities. Scottish Natural Heritage (1997) published a review of long distance routes in Scotland which included detailed recommendations about management, funding and promotion. The Scottish Office (1996b) published a guide to managing recreational water use.

Miscellaneous Issues

The Countryside Commission (1995p) published jointly with the Sports Council a report by Oxford Brookes University on good practice in managing sport and active recreation in the countryside. This looked at 12 examples of good practice in 19 forms of sport/active recreation in eight types of conservation designation. It concluded that conflict between conservation and active sport/recreation can, in almost all cases, be prevented or resolved by good planning and management practice. In particular, voluntary agreements can be effective, but problems often arise from activities by non-club members who are not bound by or are unaware of codes of practice. For the future, the report set out six key considerations which included: acquiring a thorough knowledge of the baseline environmental conditions; encouraging participatory management; and involving local people.

The Countryside Commission (1995r) issued an information pack which enables countryside recreation site managers to assess the quality of welcome offered to visitors at their sites. The assessment is based on site care, promotion, signing, accessibility and safety and recognises the importance of a managed country site for many people. In a similar vein the Countryside Commission (1996j) published guidelines on how to carry out market research for recreation. This emphasised that finding out what visitors want from recreation is an essential first step when planning improvements or new facilities.

Schemes

The Country Landowners Association launched a new access initiative entitled ' Access 2000' in May 1996, in order to build on the 540,000 hectares of access negotiated under schemes like Countryside Stewardship between 1990 and 1995. According to McDougal (1996c) this stressed that the way to increase access was via voluntary agreements between an alliance of countryside organisations, farmers and landowners. Some commentators saw this as an attempt to ward off the Labour Party's interest in a formal right to roam strategy. The Ramblers Association argued that voluntary agreements had not worked in the past and continued to call for a right to roam policy.

The Countryside Commission launched a campaign to help 1,000 communities to create local open spaces that would remain accessible to the public in perpetuity in May 1995. Originally it had been hoped to raise £45 million from the National Lottery, but only £10 million was granted in the summer of 1996. Therefore the Commission invited bids from local communities in 1996 which it hoped would lead to around 250 Millennium Village Greens being created by the year 2000. In 1994 and 1995 the Commission was given approval to create and seek funding for a national trail along the line of Hadrian's Wall and for a Pennine Bridleway trail. A national network of cycle trails being developed by the charity Sustrans was given a major boost in 1995 when it gained a grant worth up to £42.5 million from the National Lottery. It was expected that it would take till 2005 to complete the 6,500 mile network at a total cost of £183 million.

The National Trust issued the findings of an Access Review Working Party. This set out three principles. **Principle 1** The duty and primary purpose of the Trust in the countryside is to promote permanent preservation. It will regard access as a fundamental way of providing this benefit and as a principal purpose. **Principle 2** The National Trust Acts establish the Trust's responsibilities for conservation. If serious conflict arises, conservation will take precedence over access. **Principle 3** The Trust will ensure that the countryside retains characteristics which afford the widest range of experiences, and will enable people to enjoy access to its properties.

In Scotland, Scottish Natural Heritage launched a major new initiative in 1996 to improve local access for walkers, horse riders and cyclists around lowland villages, towns and cities. Entitled 'Paths for All-the Scottish Way', the initiative was to be implemented by a Paths for All Partnership supported by 15 organisations and staffed by four people. The immediate aim was to improve access to the local countryside, with the longer term aim of connecting communities into longer routes.

Studies and Commentaries

The regular surveys of day visits to the countryside and the coast by the Countryside Commission, based on household survey, continued to show the popularity and economic significance of such day trips, with over 1 billion visits being made and nearly £6 billion spent. The first survey of visitors to all 10 National Parks, the Norfolk Broads and the New Forest was carried out in 1994 (Countryside Commission, 1997b). 68,000 completed questionnaires revealed that good scenery and landscape were the main attractions followed by fresh air and peace and quiet. Less than 20% had any complaints, but 24% of visitors to the Lake District said overcrowding was a problem. Average spending per head of £6.90 was higher than the average spend of £5.17 to the countryside in general.

A survey of the 105,000 mile Rights of Way network by the Countryside Commission (1996k) found it to be in a better condition than for decades. Well over three-quarters were open for use, with 90% usable compared to 68% in 1988, and there was a better than 50% chance of successfully completing a two-mile walk without meeting an impassable obstacle, compared to a 33% chance in 1988. Between 1988 and 1996 the Commission had spent £16 million on improving Rights of Way but acknowledged that it was unlikely to meet its target of having every path legally defined and properly maintained by the year 2000. The Ramblers Association in contrast accused the Commission of complacency and argued that it should reallocate funds to give Rights of Way a higher priority, that it should encourage the more dilatory local authorities to use their enforcement powers to remove obstacles and that it should lobby farming organisations to lean on their members to make sure they obey the law. In more detail Curry *et al* (1996) published a review of operation of Public Paths Orders.

Watkins (1996) in an edited book examined the roots and results of the conflict between those seeking greater access and those seeking to reduce access to the countryside through a variety of approaches, whether polemical or theoretical. The overall premise is that a sea change in access to the countryside is about to occur arising from the need to find new uses for agricultural land, social change in rural areas and the growing demand for consumption of the rural heritage.

A holistic and challenging commentary has been provided by Clark *et al* (1994) in a research report and background papers commissioned by the CPRE. According to Owen (1995) these challenge the conventional wisdom that rural recreation and tourism is desirable, that different land uses can be balanced by land use planning, and that conflicts about leisure are not major causes of concern and can be negotiated or managed satisfactorily. Instead, the report argues that people do not have clear cut and unitary attitudes to the countryside, but that their responses to the same issues vary markedly with the cultural and political contexts in which they are asked questions. It is concluded that Britain faces a cultural crisis crystallising around conflicts about leisure and tourism and that there is a need for new patterns of strategic reflection and action.

More prosaically, Bell (1997) has provided a well-illustrated manual on the design aspects of providing outdoor recreation facilities, ranging from providing information and interpretation to detailed aspects such as wildlife viewing. The book is based on North American experience and examples, but the principles are universal.

6 Social and Economic Issues

Actual and Proposed Legislative Change

Economic and Social Activity

The *Local Government and Rating Act 1997* implemented a White Paper commitment to reduce substantially the rates bills of rural shops, post offices and other rural businesses in Great Britain. Local authorities were given the power to grant 50% relief from business rates to general stores or post offices in settlements of 3,000 people or less. They were also given discretion to waive the remaining 50% and to provide 100% rate relief for any other rural businesses which they consider benefit the local community.

The Rural Challenge initiative reached its fourth round of bidding by 1997. In 1996 the six winners of the £1 million per annum prizes included: a £1.5 million rural youth project in Somerset and a £4.3 million project to restore part of the Ashby canal at Measham in Leicestershire. A 1997 report on the first two sets of winners by the Rural Development Commission argued that winning bids have a clear focus, a sound package of funding from other sources, a respect for the environment and an understanding of how the work will be sustained in the long term. In terms of completing the projects they must be flexible, closely managed and involve the community.

Significant funding for rural areas continued to come from the European Union. In November 1994, 225 million ECU were made available for agricultural projects between 1994-99 under Objective 5a under which the EU contribution is 25% of costs. In December 1994, 550 million ECU was made available for rural development projects in the English Northern Uplands, Southwest England, Wales and the Scottish Borders under Objective 5b. This fourth package of structural funds brought the total funding for 1994-99 under Objective 5b to over one billion ECU, with 514 million ECU for Southwest England, 483 million ECU for Wales and 82 million ECU for the Scottish Borders. In August 1995 46 million ECU was made available under the LEADER II community initiative programme provided UK authorities matched the EU funds. England was allocated 25 million ECU in Objective 5b areas, Scotland 8 million ECU in Objective 5b areas and 12 million ECU in the Highlands and Islands.

Highlands and Islands Enterprise issued a *Strategy for Enterprise Development* in 1996 which pursued three strategic objectives: growing businesses, developing people and strengthening communities in four key sectors: food and drink; manufacturing; tourism; and knowledge, information and telecommunications. The success of previous strategies are demonstrated by the facts that the area has experienced population growth of 50,000 people since the 1960s to return the population to the level of the 1920s.

Housing

The *Housing Act 1996* gave more housing association tenants the right to buy their home at a discount and provided for a register of social landlords to be kept by the Housing Corporation. Right to buy disposals were however restricted in three areas: National Parks; AONBs and designated rural areas under the 1985 Housing Act. In these areas subsequent sales are restricted to people who have lived or worked in the area for at least three years.

A White Paper on housing in England and Wales (Cm 2901, 1995) endorsed the long term movement of people into the owner-occupied sector and hoped to see a further 1.5 million in owner-occupied housing by 2005. In terms of housing's relationship to the environment the White Paper wanted: better use made of existing buildings and empty houses; more development within built up areas with a target of 50% of new housing on brownfield land; new housing in villages only where it would help to sustain the local economy and facilities and thus reduce the need to travel, while elsewhere some villages might have reached their natural capacity.

The *Local Government and Rating Bill 1997* (see above) gave English parishes and Welsh community councils powers to promote community transport. Under the Act, parishes can: set up car-sharing schemes; make grants for bus services; arrange concessionary taxi fares; publicise local transport services; and fund traffic calming. The *Road Traffic Reduction Act 1997* placed a duty on local traffic authorities to specify targets for a reduction of local road traffic in their area, or a reduction in the rate of growth.

The Government published a Transport Green Paper (Cm 3234, 1996) which also acted as a response to the Royal Commission on Environmental Pollution's 1994 report on *Transport and the Environment* (see below). The main message of the Paper was that current trends are unsustainable both environmentally and economically. It thus marked the end of the 'predict and provide' mentality which had dominated transport policy based on building new roads for several decades. However, while the Paper was welcomed for its clear diagnosis of the illness it was criticised for failing to provide either new or tough enough prescriptions, except for broad exhortations in favour of public rather than private transport.

The paper recognised the special transport needs of rural areas and proposed the promotion of present schemes including: recent cash increases for the Rural Transport Development Fund, encouraging flexible and innovative community-based services and reducing the impact of traffic in villages and sensitive areas by traffic calming.

In the meantime, the Rural Development Commission announced a Rural Transport Initiative in 1996. Funded with £90,000 over 1996-99 it aimed to: develop a resource pack to help communities assess their transport needs and ways of meeting them; and a programme of training for community and voluntary transport operators.

In Scotland, a Green Paper on transport (Cm 3565, 1997) was more friendly to road building and car use, reflecting the greater rurality of Scotland, when it stated: 'In the Government's view it is generally inappropriate and undesirable to seek to manage traffic growth in rural areas, given the suitability of car travel'. Earlier, the Scottish Office had published a set of consultation papers in 1995 on

A Balanced Approach to Rural Road Design and *Rural Road Hierarchy and Lorry Routeing* as part of a long term process which began in 1992 in order to achieve the most appropriate and consistent standards in planning, developing and maintaining rural roads.

In the energy field, the fourth non-fossil fuel obligation, in February 1997, obliged the regional electricity companies to place contracts resulting in some 843 megawatts of new renewable generating capacity between 1997 and 2016, representing enough power for more than a million homes. Some 40 per cent of the total was earmarked for wind energy projects. The fourth obligation noted that there was real convergence between renewable energy and the market rate for electricity. However, the CPRE complained that subsidising wind energy flew in the face of public concern about their landscape impact. The third non-fossil fuel obligation in January 1995 had called for 626 megawattts in England and Wales between 1995 and 2014. The jump from 626 to 843 megawatts between 1995 and 1997 demonstrated the growing importance of this sector.

In Wales, the battle had already been joined by a Welsh Affairs Committee report on Wind Energy in 1994, which had called for a presumption against wind farms either in or visible from designated landscape areas. In his reply, the Secretary of State for Wales (Cm 2694, 1994) rejected this proposal because it would effectively preclude wind energy production from the greater part of Wales. Instead, the response argued that it would be more appropriate for local plans to show the circumstances in which wind development proposals might be permitted even if these were very limited.

Agencies

The Rural Development Commission continued to fund rural employment initiatives. For example, in 1995-96, the Commission spent more than £13 million on rural development programmes, made 13,800 advisory visits to small rural firms and approved grants for 218 village halls, as well as the Rural Challenge initiative outlined above. This was similar to the £3 million spent on support for voluntary activities in rural areas in 1994-95, alongside: nearly

£7 million invested in industrial sites, providing space for 800 jobs; more than £3.3 million spent on renovating premises for 460 firms providing space for 1,600 jobs; and 16,000 advisory visits made to individual rural businesses.

In Scotland, Highlands and Islands Enterprise reflected on the first five years of their Network programme. The 1995-96 Annual Report noted that the Network had assisted over 7,000 business developments, that Network funding of around £100 million had unlocked private sector investment of nearly £270 million and that some 15,188 jobs at an average cost to the taxpayer of £2,780 per job had been created. In the more traditional funding role 3,667 jobs were created in 1995-96, over 6,000 people were assisted with training programmes and 493 Community Action Grants were approved.

In Wales, the Development Board for Wales published *Rural Wales: A Challenge of Change* in 1997 which expressed concern that rural Wales had become dangerously dependent on European funding worth £200 million a year in agricultural aid. In addition, most of the Board's construction programme, including factory premises worth £750,000, in 1996 were funded by Brussels. However, if funds could be maintained then current successes could be continued. For example, in 1996-97, 1500 new jobs and £32 million of investment from the private sector had been achieved. Because agriculture and tourism account for 25 and 9 per cent of all jobs in rural Wales it was essential to keep attracting manufacturing jobs when the Board had helped to increase manufacturing jobs in rural Wales by 14% compared to a decline of 24% in Britain.

The Department of Transport issued three annual reports (Cm 2803, 1995; Cm 3206, 1996; Cm 3604, 1997) none of which contained any specifically rural issues. They did however reflect the growing emphasis on sustainability as transport policy moved away from private vehicle provision and road building. For example, spending on roads was set to fall from £2,609 million in 1993-94 to £1,757 in 1999-2000.

The Office of Population Censuses and Surveys merged with the Central Statistical Office to form the Office for National Statistics on 1 April 1996. The merger had most effect in England and Wales.

Scotland continued to have its own statistical departments but co-operated in the production of UK statistics.

Studies

General and Population

Cloke and Little (1997) have edited a book about a range of neglected groups, mainly, women, the old and the young under the generic title of '*Contested countryside cultures*'. The book contains 15 chapters, of which ten take a synthetic approach in that they review existing work, and of which five report on research findings. Apart from general considerations of rurality the following themes are covered: counterurbanisation; horror movies; lesbian communities; village life for married women; employment; childhood; old age; ethnicity; and craft production. Without exception each chapter is a good read and some are entertaining too. One of the most acute, but tantalislingly short observations is made in the concluding chapter when it is observed that: '*Struggles over 'control' of rural space are also struggles-at different levels-over control of the 'meaning' of rurality and of the values that are seen as being represented by rural society and culture.*' (p.279). In this sense, the title of the book is misleading since it implies a study of contestation, but each chapter examines its own group rather than the interaction between them and the 'powerful'. Readers will thus find little here to advance their knowledge about how groups contest rural space via the planning process. (Based on a review in *International Planning Studies*.)

A more traditional view of rural disadvantage is provided by Shucksmith *et al* (1997) who were commissioned by the Rural Development Commission to undertake a review of the existing literature. The review found that there are few major studies of rural disadvantage. Those that do exist tend to provide a picture of disadvantage at one point in time and consequently, little is known about the dynamic processes by which people may move into or out of disadvantage. The review also notes that rural areas are undergoing major changes but that little is known about the uneven impacts of these changes in different rural areas and on different groups. However, the review does argue that we know more about some specific issues, such as low incomes and access to services, but even here there are some significant gaps in knowledge.

Another source of disadvantage, different provisions of public expenditure, has been examined by the Rural Development Commission (1996b). They examined how public funds are distributed to local authorities and district health authorities by standard formulas which take into account local factors. These Standard Spending Assessments are intended to provide a standard level of service provision across the country, but this was not found to be the case, with shire counties receiving an average £648 per resident compared to £756 for metropolitan areas and £1,154 for inner London. The report suggests three reasons for these differences. First, little account is taken of the costs of rurality, second, some of the measures reflect urban rather than rural characteristics, and third, the measures overestimate higher costs incurred in cities.

In Scotland, Shucksmith *et al* (1996a) have written up the findings of a 2 year study into how people live in rural Scotland today. The study interviewed 600 rural people between 1993 and 1994 about how they live and what they think about their lives and circumstances. This revealed that poverty is widespread but that people's subjective assessment of their position is at odds with objective definitions. Take-up of benefits are low, raising issues of overcoming resistance to entitlement and of the invisibility of the excluded. Housing was a pervasive problem, as were restricted work options. Services were also a matter of concern, especially, the cost of essential private transport. Most worryingly, rapid change and an ability to influence events were central themes and rural people feel disempowered, distant from decisions which affect their lives and at the mercy of events.

Other measures of rural life in Scotland based on census and other published statistics have been provided by Williams *et al* (1996b) in their update of 1992 to 1995 data and by a Scottish Office (1996c) study of four rural communities.

opportunities. The research concluded that constraints on land supply rather than the nature of demand have been the main influence over the level of housing opportunities.

Accordingly, the Housing Corporation (1995) argued for more 'exceptions' to be made for affordable housing schemes by local planning authorities in order to boost the inadequate total of 7,000 housing association homes built between 1992 and 1995. In a study of how local authorities specify housing tenure when granting planning permission on community grounds the Rural Development Commission (1995a) found that planners caused many problems and so called for a review of the system. Nonetheless, the main problem according to a survey by Bramley (1995) for the Rural Development Commission remains one of low rural incomes which revealed that in at least 36 rural districts over 40% of new households cannot afford to buy. Research in the specific case of the Green Belts by Elson *et al* (1996) found that only 49 affordable schemes for 414 units had been approved in the 146 local authorities they surveyed and that only half the authorities had considered using the 'exceptions' policy in PPG3.

Employment

Allanson and Whitby (1997) have edited a book on *The Rural Economy* which aims to summarise some of the findings of the major publicly funded research programmes into rural employment issues of recent years. It is not however consistent in this aim and this when combined with the notorious eclecticism of edited books makes this book a rather partial view of the wider countryside scene, although sustainability does emerge as an accidental if unstated aim of the book. Nonetheless, there are many good points in the book. Notably, an elegant exposition of the evolution of rural policy by the editors, a compelling account of how different value systems could be used in policy making by Harvey, a sound text book analysis of methods for valuing environmental goods by Garrod, an introduction to the issue of marketing rural food products by Ritson and Kuznesof which will surely be the issue of the next decade, a neat analysis of winners and losers from rural policy by Whitby, and finally, a perceptive analysis of how the three prime objectives of rural policy:

efficiency; equity; and sustainability have been met in post war rural Britain. (A longer version of this review by the author appeared in *Town and Country Planning.*)

In terms of implementing policies a study by Oxford Brookes University for the Department of the Environment (1995s) revealed that planners were implementing policies in PPG7 which called for a more relaxed approach to rural diversification on farms. The study was accompanied by a good practice guide (Environment, Department of, 1995t). For more detail see Chapter Three.

Examples of how communities could help themselves to achieve economic regeneration have been provided by Minay (1995) who identified 100 rural community schemes, for example the Lockinge Estate project in Oxfordshire. In a review of the top down approach, the House of Commons Trade and Industry Committee (1995) criticised the bewildering profusion of schemes on offer and the multitude of agencies administering and advising on them and argued that substantial savings could be made by streamlining the system. Nonetheless, they believed that regional policy is an effective way of creating jobs and promoting competitiveness at a reasonable cost. With regard to rural areas the Committee argued that they have special problems and that there is a need for a comprehensive review of these problems and the role of regional and other policies in tackling them. It also called for a review of regional aid policy and in particular for a revival of regional strategies so that the funds could be co-ordinated for maximum effectiveness. In Scotland, the Scottish Office (1996d and 1997) published a scoping study of rural development and discussed new ideas in rural development based on community involvement.

The Rural Development Commission published a report on the impact of rural tourism by Seagull Quince Wicksteed (1996) which revealed that medium sized settlements (1,400-3,250 people) benefited most. The study based on 11 different types of settlement found that basic shops, such as food shops, were significantly enhanced, other businesses became more viable, services were more sustainable and that there were positive contributions to community life. One year earlier the Rural Development Commission (1995b) had published a policy for rural tourism which contained five objectives including: support for the extension of seasons and the geographical spread of tourism; encouraging high-quality tourism facilities and services; and support for efforts to halt tourist products and areas in decline.

The Rural Development Commission also published a report on the impact of the peace dividend on rural England prepared by the Economists Advisory Group and ECOTEC (1996). This revealed that although there had been significant direct and indirect job losses as defence bases had been closed, there did not appear to have been a major employment crisis.

Advocacy

Transport and Energy

The policy changes and studies outlined earlier in the chapter were also endorsed by a number of advocacy reports. For example, Secrett (1996) published a paper on moving towards a sustainable transport sector which advocated setting targets for reducing travel and to develop transport tax policies based on the principles of ecological tax reform, namely, moving the burden of taxation away from 'social' goods like labour, income and capital, and onto environmental 'bads' like pollution and waste.

The Rural Development Commission (1996c) made a plea that the transport debate should not however be dominated by unilateral or by urban based thinking and that rural transport should be treated separately, because rural people travel 50% further and their journeys are 40% longer. A car is thus a necessity for most rural people. The Commission called for a three pronged approach. First, that policy should recognise the special needs of rural areas and differentiate between the two. Second, to reduce the need to travel by improving the availability of jobs and services in small towns and villages. Third, to improve public and community transport by for example using paid drivers in community bus schemes. To help achieve better community and voluntary transport the Rural Development Commission (1996d) published a series of recommendations which it began to test via the Rural Transport Initiative already outlined above.

The Countryside Commission (1995s) also criticised the urban dominated view of transport, but this time in the way that roads were built in ways very unsympathetic to rural landscapes. Accordingly it proposed an approach based on a hierarchy of roads and for local roads to be restored to pre-suburban standards. This would include the restoration of hedgerows and stone walls where possible and the revival of black and white finger posts.

The National Trust has long been a critic of road improvement schemes and has indeed been the victim in some cases. Accordingly, they have published a booklet by Liniado (1996) which examines cultural attitudes to the car and how the National Trust by limiting car parking spaces and offering discounts for visitors who arrive by public transport can help to weaken the grip of the motor car in popular imagination.

A report by David Davies Associates (1996) for a consortium of pressure groups however found that local authority spending and bidding for funds was still dominated by road programmes aimed at benefiting car and lorry users. Accordingly, they asked Ministers to give local authorities adequate tools for demand management and stronger encouragement and funding for rural transport packages.

Central government guidance was also criticised by the Council for the Protection of Rural England (1997) notably PPG6 and PPG13. The Council proposed a tax on business and retail car parking and regional strategies to examine alternative transport solutions and regional parking strategies. The Council also published forecasts of how rural roads would experience massive increases in use if nothing was done. For example, traffic was forecast to grow by 300% by 2025 in Dorset, Sussex, Nottinghamshire and Cambridgeshire. The Council had already contributed to the great transport debate (1995f), endorsed the Countryside Commission's critique of rural road suburbanisation (1995g) and called for a new environmentally sensitive approach to managing the road network (1995h).

In contrast the Automobile Association (1997) while recognising the problems raised by growing ownership and use of cars called for more road investment and for the completion of at least 300 of Britain's 600 unbuilt bypasses, widening 500 miles of motorways and finishing 'missing' trunk road links by 2005.

Employment

The Rural Development Commission issued a number of documents concerning employment. First, they published (Rural, 1995c) a contribution on *Rural Economic Activity* to the rural White Paper which aimed to influence policy on small rural firms, on recycling existing rural buildings for new businesses and on the need for assistance to priority and less developed areas. Second, they issued (Rural, 1995d) a policy statement which called for a more creative and proactive role for the planning system in which villages would become places where people lived and worked. Third, they published (Rural, 1996e) a guide to how old buildings can create new opportunities for redundant buildings by converting them to new uses which helps to save the nation's built heritage.

Housing

A report by Oldfield King Planning (1996) for the Rural Development Commission found that the majority of local authority strategies for rural housing had failed, in spite of DOE guidance, to identify specific rural housing needs and to undertake a housing needs survey which distinguished rural areas. Accordingly, the report recommended the inclusion of a specific chapter on rural housing in DOE guidance, a good practice guide to preparing rural housing strategies, further clarification of planning policy guidance on providing sites for housing in smaller settlements, and adopting local measures to assist the provision of affordable housing such as 'options land banks' and 'rural housing enablers'.

References

Abel, C. (1995) Free Market: Growers now face the great unknown, *Farmers Weekly*, 27 October, pp. 54-55.

Adams, W. (1996) *Future Nature: A vision for conservation*, Earthscan.

Advisory Panel on Standards for the Planning Inspectorate Agency (1995) *Second Report*, Department of the Environment.

---- (1996) *Third Report*, Department of the Environment.

Ageyman, J. and Evans, B. (eds) (1994) *Local Environmental Policies and Strategies*, Longman.

Agriculture, Fisheries and Food, Ministry of (1995a) *European Agriculture - The Case for Radical Reform*, The Ministry.

---- (1995b) *Environmental Land Management Schemes in England - A Consultation Document*, The Ministry.

Alexander, D. (1995) Whitehall clears decks on land management regime, *Planning*, 13 October, pp. 24-25.

Allan, J. (1995) A slow response to woodland set-aside, *Farmers Weekly*, 20 October, p. 59.

Allanson, P. and Whitby, M. (eds) (1997) *The Rural Economy and the British Countryside*, Earthscan.

Association of Metropolitan Authorities (1996a) *Environmental Manifesto for Local Government*, The Association.

---- (1996b) *The future of the planning system - A consultation paper*, The Association.

Aston Business School (1997) *Evaluation of Rural Action for the Environment*, Aston University.

Audit Commission (1997) *Representing the People: The role of councillors*, The Commission.

Baldock, D., Bishop, K., Mitchell, K. and Phillips, A. (1996) *Growing Greener: Sustainable Agriculture in the UK*, CPRE and WWF.

Banister, D. (1995) Transport and the Environment: A Review Article, *Town Planning Review*, 66, pp. 453-8.

Banks, S. (1994) *Practical Planning: Appeals and Inquiries*, Longman.
Barkham, J. (1996) Government response to the UK Steering Group Report on Biodiversity, *Ecos*, 17 (1), pp. 92-93.
Barrow, C. (1995) *Developing the Environment: Problems and Management*, Longman.
Bartram, H. (1997) Hedgerows, *Ecos*, 18(1), p. 95.
Bell, S. (1997) *Design for Outdoor Recreation*, E & FN Spon.
Bibby, P. and Shepherd, J. (1996) *Urbanisation in England: Projections 1991-2016*, HMSO.
Booth, P. (1996) *Controlling development: Certainty and discretion in Europe, the USA and Hong Kong*, UCL Press.
Bough, J. (1996) Review of 'Nature Conservation and Countryside Law', *Ecos*, 17(3/4), pp. 91-92.
Bowers, J. (1996) Review of 'Indicators of Sustainable Development for the United Kingdom', *Ecos*, 17(2), pp. 93-95.
Bramley, G. (1995) *Rural Incomes and Housing Affordability*, Rural Development Commission.
Bramley, G., Bartlett, W. and Lambert, C. (1995) *Planning the market and private house-building*, UCL Press.
Bramley, G. and Watkins, C. (1996) *Circular projections*, Council for the Protection of Rural England.
Brindley, T., Rydin, Y. and Stoker, G. (1996) *Remaking planning: The Politics of Urban Change*, Routledge.
British Government Panel on Sustainable Development (1995) *First Report*, DOE Publications.
---- (1996) *Second Report*, DOE Publications.
---- (1997) *Third Report*, DOE Publications.
Brown, S. (1996) *Roots 1996: Proceedings of the 1996 Rural Practice Research Conference*, RICS.
Bryden, J. and Mather, A. (1996) The 'rural' White Paper-Rural Scotland, *Scottish Geographical Magazine*, 112, pp. 114-116.
Buckingham, S. and Evans, B. (eds) *Environmental planning and sustainability*, Wiley.
Bullen, P. (1995a) Labour promises a better countryside-but at a price, *Farmers Weekly*, 21 April, p. 12.
---- (1995b) CLA sets out rural vision of the future, *Farmers Weekly*, 17 February, p. 12.

---- (1995c) Tenancy Act won't mean a 'new dawn' for young, *Farmers Weekly*, 19 May, p.16.
---- (1995d) Cross-compliance finds little backing within EU, *Farmers Weekly*, 3 November, p. 16.
Burton, I. (1994) Cairngorms partners to take over mountain management, *Planning*, 18 November, p. 10.
Burton, T. (1995) The Environment Act 1995: blessing or bane? *Ecos*, 16 (3/4), pp. 3-5.
Callander, R. (1995) *Forests and People in Rural Scotland: A discussion paper*, Rural Forum Scotland.
Candy, P. (1994) Organic window opens slowly on rural scene, *Planning*, 9 December, p. 10.
Carmona, M. (1996) The recently issued revised draft of PPG1, *Town and Country Planning*, September, pp. 240-243.
Centre for Environment and Land Tenure Studies (1996) *Countryside Stewardship payments, farm incomes and the capital value of farm land*, Reading University.
Cheesley, A. (1995) CLA-switch to rural policy, *Farmers Weekly*, 28 April, p. 7.
Cherry, G. (1997) *Town Planning in Britain since 1900*, Blackwell.
Cherry, G. and Rogers, A. (1996) *Rural Change and Planning: England and Wales in the Twentieth Century*, E & FN Spon.
Clark, G., Darrall, J., Grove-White, P., MacNaghton, P. and Urry, J. (1994) *Leisure, Culture and the English Countryside*, CPRE.
Cloke, P. and Little, J. (eds) (1997) *Contested countryside cultures: otherness, marginalisation and rurality*, Routledge.
Cm 2674 (1994) *Transport and the Environment: 18th Report of the Royal Commission on Environmental Pollution*, HMSO.
Cm 2694 (1994) *The Government's response to the Welsh Affairs Committee Report on Wind Energy*, HMSO.
Cm 2750 (1994) *Next Steps Agencies in Government: Review 1994*, HMSO.
Cm 2767 (1995) *The Government's response to the Environment Committee's report on Shopping Centres*, HMSO.
Cm 2803 (1995) *1995 report by MAFF and the Intervention Board*, HMSO.
Cm 2807 (1995) *1995 report by the Department of the Environment*, HMSO.

Cm 2814 (1995) *1995 Report by Secretary of State for Scotland and the Forestry Commission*, HMSO.
Cm 2815 (1995) *1995 report by the Welsh Office*, HMSO.
Cm 2822 (1997) *This Common Inheritance: UK annual report 1995*, HMSO.
Cm 2901 (1995) *Our Future Homes*, HMSO.
Cm 3016 (1995) *Rural England*, HMSO.
Cm 3018 (1995) *Government response to the House Of Lords Select Committee Report on Sustainable Development*, HMSO.
Cm 3041 (1995) *Rural Scotland*, HMSO.
Cm 3164 (1996) *Next Steps Agencies in government: Review 1995*, HMSO.
Cm 3179 (1996) *Spending public money: governance and audit issues*, HMSO.
Cm 3180 (1996) *A working countryside for Wales*, HMSO.
Cm 3188 (1997) *This Common Inheritance: UK annual report 1996*, HMSO.
Cm 3204 (1996) *1996 report by MAFF and the Intervention Board*, HMSO.
Cm 3207 (1996) *1996 report by the Department of the Environment*, HMSO.
Cm 3214 (1996) *1996 Report by Secretary of State for Scotland and the Forestry Commission*, HMSO.
Cm 3215 (1996) *1996 report by the Welsh Office*, HMSO.
Cm 3234 (1996) *Transport: The Way Forward*, HMSO.
Cm 3259 (1996) *Housing Need: Government response to House of Commons Environment Committee Report*, HMSO.
Cm 3260 (1996) *Government response to the UK Steering Group on Biodiversity*, HMSO.
Cm 3267 (1996) *Protecting the built heritage: A consultation document about Scotland's built heritage*, HMSO.
Cm 3323 (1997) *United Kingdom National environmental health action plan*, HMSO.
Cm 3444 (1996) *Rural England: A report on progress made since the 1995 White Paper*, HMSO.
Cm 3471 (1996) *Household growth: Where shall we live?*, HMSO.

Cm 3548 (1997) *Scottish Office reply to House of Commons Scottish Affairs Committee inquiry into the future of Scottish agriculture*, HMSO.
Cm 3556 (1997) *This Common Inheritance: UK annual report 1997*, HMSO.
Cm 3557 (1997) *The governance of public bodies: a progress report*, HMSO.
Cm 3562 (1997) *Government response to Environment Committee report on water conservation and supply*, HMSO.
Cm 3565 (1997) *Keeping Scotland moving*, HMSO.
Cm 3579 (1994) *Next Steps Agencies in government: Review 1996*, HMSO.
Cm 3604 (1997) *1997 report by MAFF and the Intervention Board*, HMSO.
Cm 3607 (1995) *1997 report by the Department of the Environment*, HMSO.
Cm 3614 (1997) *1997 Report by Secretary of State for Scotland and the Forestry Commission*, HMSO.
Cm 3615 (1995) *1997 report by the Welsh Office*, HMSO.
Council for the Protection of Rural England (1994) *The Housing Numbers Game*, The Council.
---- (1995a) *Water Water Everywhere*, The Council.
---- (1995b) *Tranquil Area Maps of England*, The Council.
---- (1995c) *Index of Planning Policy Guidance Notes*, The Council.
---- (1995d) *Local Attraction*, The Council.
---- (1995e) *Local Influence: Increasing local involvement in the development of green farming schemes*, The Council.
---- (1995f) *The Great Transport Debate*, The Council.
---- (1995g) *Changing Lanes*, The Council.
---- (1995h) *The End of Hierarchy: A new perspective on managing the road network*, The Council.
---- (1996a) *70 years of achievement*, The Council.
---- (1996b) *Testing! Testing! Regional Housing Requirements in Development Plans*, The Council.
---- (1996c) *Campaigners Guide to Minerals*, The Council.
---- (1997) *Planning more to travel less*, The Council.
Country Landowners Association (1994) *Focus on the CAP*, The Association.

---- (1995a) *Towards a Rural Policy: a vision for the 21st century*, The Association.
---- (1995b) *European Rural Policy*, The Association.
Countryside Commission (1994a) *Working for the countryside 1994/95: CCP 476*, The Commission.
---- (1994b) *The High Weald: Exploring the landscape of the Area of Outstanding Natural Beauty*, CCP466, The Commission.
---- (1994c) *The Cannock Chase Landscape*, CCP469, The Commission.
---- (1994d) *Military training in the Northumberland National Park: Advice to the Ministry of Defence*, CCP471, The Commission.
---- (1994e) *Design in the countryside experiments*, CCP473, The Commission.
---- (1994f) *The National Forest: The Strategy*, CCP468, The Commission.
---- (1994g) *The Rights of Way Act 1990: Guidance Notes for Farmers*, CCP299, The Commission.
---- (1995a) *Quality of Countryside: Quality of Life-The Countryside Commission's prospectus into the next century, CCP470*, The Commission.
---- (1995b) *Annual report 1994-95: CCP480*, The Commission.
---- (1995c) *Tamar Valley Landscape: A Landscape Assessment*, The Commission.
---- (1995d) *The Cannock Chase Landscape: An introduction to the landscape of the Chase*, CCX33, The Commission.
---- (1995e) *The Cranborne Chase and West Wiltshire Downs Landscape*, CCP465, The Commission.
---- (1995f) *The Solway Coast Landscape*, CCP478, The Commission.
---- (1995g) *The Kent Downs Landscape*, CCP479, The Commission.
---- (1995h) *Heritage Coasts: A guide for councillors and officers*, CCP475, The Commission.
---- (1995i) *Climate change, acidification and ozone: Potential impacts on the English countryside*, CCP458, The Commission.
---- (1995j) *Grants and Payment Schemes*, CCP422, The Commission.

---- (1995k) *Growing in Confidence: Understanding people's perceptions of urban fringe woodlands*, CCP457, The Commission.
---- (1995l) *Sustainable Rural Tourism: Opportunities for local action*, CCP483, The Commission with Department of National Heritage, Rural Development Commission and English Tourist Board.
---- (1995m) *Principles of Sustainable Rural Tourism: Opportunities for local action*, The Commission with Department of National Heritage, Rural Development Commission and English Tourist Board.
---- (1995n) *Education Access Initiative: Information Booklet for Land Managers*, CCP464, The Commission.
---- (1995o) *The Way Forward: Options for the Future Management of England's National Trails*, CCP437, The Commission.
---- (1995p) *Good practice in the planning and management of sport and active recreation in the countryside*, The Commission and the Sports Council.
---- (1995r) *Visitor Welcome Initiative*, CCP476, The Commission.
---- (1995s) *Roads in the countryside*, CCP459, The Commission.
---- (1996a) *A living countryside: CCP492-Our strategy for the next ten years; CCP493-A summary; and CCP494-Getting started-a checklist for partners of the Commission's plans for the four years beginning April 1996*, The Commission.
---- (1996b) *Annual report for 1995-96: CCP499*, The Commission.
---- (1996c) *The Norfolk Coast Landscape*, CCP486, The Commission.
---- (1996d) *The Wye Valley Landscape*, CCP487, The Commission.
---- (1996e) *The Howardian Hills Landscape*, CCP474, The Commission.
---- (1996f) *The Landscape of the Sussex Downs*, CCP495, The Commission.
---- (1996g) *England's countryside-The role of the planning system*, The Commission.
---- (1996h) *Village design-making local character work in new development*, CCP501, The Commission.
---- (1996i) *Managing Rights of Way: A Discussion Paper*, CCP506, The Commission.

---- (1996j) *Market Research for Countryside Recreation*, The Commission.
---- (1996k) *Second National Rights of Way Condition Survey 1993-94*, CCP504 and CCP505 (summary leaflet), The Commission.
---- (1997a) *Village Design: making local character count in new development*, CCP501 and *Countryside Design Summaries*, CCP502, The Commission.
---- (1997b) *Visitors to National Parks-Summary of the 1994 Survey Findings*, CCP503, The Commission.
---- English Heritage and English Nature (1996) *Ideas into action for Local Agenda 21*, The Commission.
Countryside Council for Wales (1996) *The Welsh Landscape- Our Inheritance and its Future Protection and Enhancement*, The Council.
Culley, G. (1995) A milestone on the nature trail, *Planning Week*, 19 January, pp. 16-17.
Curry, N. and Owen, S. (eds) (1996) *Changing Rural Policy in Britain*, The Countryside and Community Press.
Curry, N., Ravenscroft, N. and Edwards, D. (1996) *Review of Local Authorities Charging for the Making of Public Path Orders*, Countryside and Community Press.
David Davies Associates (1996) *At the crossroads-Investing in sustainable transport*, Transport 2000.
Davies, R. (1996) Welsh rural affairs have Labour pledge, *Farmers Weekly*, 22 March, p. 13.
Delafons, J. (1995) Ten new unitaries, *Town and Country Planning*, October, p. 283.
---- (1996) PPG1 (Revised) Revised, *Town and Country Planning*, September, p. 226.
Denison, N. (1996) Note locks the door on rival retail plans, *Planning*, 19 July, pp. 8-9.
Dwyer, J. and Hodge, I. (1996) *Countryside in Trust: Land Management by Conservation, Recreation and Amenity Organisations*, Wiley.
Earthy, T. and Dodd, A. (1996) First step on the road to environmental appraisal, *Planning*, 15 March, p. 20.

Economists Advisory Group and ECOTEC (1996) *The Impact of the Peace Dividend on Rural England*, Rural Development Commission.

Elson, M. (1995) The Environmental Impact of Leisure Activities, *Ecos*, 16 (3/4), pp. 91-92.

Elson, M. (1996) Planning good things for the countryside?, *Town and Country Planning*, January, pp. 14-16.

Elson, M., Steenberg, C. and Mendham, N. (1996) *Green Belts and Affordable Housing: Can we have both?*, Policy Press with Joseph Rowntree Foundation.

English Heritage (1996) *Wetlands: archaeology and nature conservation*, HMSO.

English Heritage, Countryside Commission and English Nature (1996) *Conservation issues in Local Plans*, Heritage.

English Nature (1994) *Planning for Environmental Sustainability*, Nature.

---- (1995a) *Environmental Management Policy Statement*, Nature.

---- (1995b) *Rebuilding the English Countryside-Habitat fragmentation and wildlife corridor*, Nature.

---- (1996) *Impact of water abstraction on SSSIs*, Nature.

English Nature and Countryside Commission (1996) *The character of England: landscape, wildlife and natural features*, Nature.

Environment Agency (1996a) *Corporate Plan 1997-98*, The Agency.

---- (1996b) *The Environment of England and Wales-A snapshot*, The Agency.

---- (1996c) *Environmental Assessment: Scoping Handbook for Projects*, HMSO.

Environment, Department of (1994a) *PPG15: Planning and the Historic Environment*, HMSO.

---- (1994b) *Countryside 1990 Series: Volume 3- Comparison of Land Cover Definitions; Volume 4- Development of Countryside Information System; Volume 7- Processes of countryside change in Britain; Volume 8- Environmental Accounts for Land Cover; and Volume 9- Policy Review*, The Department.

---- (1994c) *PPG9: Nature Conservation*, HMSO.

---- (1994d) *PPG24: Planning and noise*, HMSO.

---- (1994e) *Circular 18/94: Gypsy sites policy and unauthorised camping*, HMSO.

---- (1994f) *Evaluation of Environmental Information for Planning Projects: A good practice guide*, HMSO.
---- (1994g) *Good Practice on the Evaluation of Environmental Information for Planning Projects*, HMSO.
---- (1995a) *Towards Sustainability*, HMSO.
---- (1995b) *A guide to risk assessment and risk management for environmental protection*, HMSO.
---- (1995c) *Circular 2/95: The voluntary eco-management and audit scheme (EMAS) for local government*, HMSO.
---- (1995d) *Water conservation Government Action*, The Department.
---- (1995e) *PPG2: Green Belts*, HMSO.
---- (1995f) *'PPG13'-A Guide to Better Practice*, HMSO.
---- (1995g) *Circular 9/95: General Development Order Consolidation*, HMSO.
---- (1995h) *Circular 10/95: Planning controls over demolition*, HMSO.
---- (1995i) *Circular 11/95: The use of conditions in planning permissions*, HMSO.
---- (1995j) *Planning permission: A guide for business*, The Department.
---- (1995k) *Policy Guidelines for the Coast*, The Department.
---- (1995l) *Index of planning guidance*, HMSO.
---- (1995m) *Attitudes to town and country planning*, HMSO.
---- (1995n) *Analysis of responses to the discussion document 'Quality in Town and Country'*, HMSO.
---- (1995o) *The efficiency and effectiveness of local plan inquiries*, HMSO.
---- (1995p) *Community involvement in planning and development processes*, HMSO.
---- (1995q) *Planning controls over agricultural and forestry development and rural building conversions*, HMSO.
---- (1995r) *Planning for rural diversification*, HMSO.
---- (1995s) *Planning for rural diversification: A good practice guide*, HMSO.
---- (1995t) *Land Use Change in England: Issue no 10, 1990*, The Department.
---- (1995u) *The Use of Article 4 Directions*, HMSO.

---- (1995v) *Derelict land prevention and the planning system*, HMSO.
---- (1995w) *Projections of households in England to 2016*, HMSO.
---- (1995x) *Circular 3/95: Permitted Development and Environmental Assessment*, HMSO.
---- (1995y) *Preparation of environmental statements for planning projects that require environmental assessment: a good practice guide*, HMSO.
---- (1995z) *Biodiversity: The UK Steering Group: Volume 1: Meeting the Rio Challenge; Volume 2: Action Plans*, HMSO.
---- (1995aa) *A classification of rural housing markets* by M. Shucksmith, HMSO.
---- (1996a) *Environmental responsibility: A review of the 1993 Toyne report by A. Ali Khan*, HMSO.
---- (1996b) *Indicators for sustainable development for the United Kingdom*, HMSO.
---- (1996c) *Environment Agency Statutory Guidance: Explanatory Document*, The Department.
---- (1996d) *Water resources and supply: agenda for action*, HMSO.
---- (1996e) *Circular 12/96: National Parks*, HMSO.
---- (1996f) *Climate change and the demand for water*, HMSO.
---- (1996g) *PPG6: Town centres and retail developments*, HMSO.
---- (1996h) *Circular 4/96: Local Government change and the planning system*, HMSO.
---- (1996i) *Circular 13/96: Planning and affordable housing*, HMSO.
---- (1996j) *Circular 15/96: Planning Appeal Procedures*, HMSO.
---- (1996k) *Sustainable Settlements and Shelter*, HMSO.
---- (1996l) *Land Use Change in England: Issue no 11, 1991*, The Department.
---- (1996m) *Changes in the quality of environmental assessments for planning projects*, HMSO.
---- (1996n) *Towards a methodology for costing biodiversity targets in the UK*, HMSO.
---- (1996o) *Reclamation of damaged land for nature conservation*, HMSO.
---- (1997a) *PPG1: General Policy and Principles*, HMSO.

---- (1997b) *PPG7: The countryside-Environmental quality and economic and social development*, HMSO.
---- (1997c) '*PPG13': Supplement: Regional Planning Guidance and Transport*, HMSO.
---- (1997d) *Circular 1/97: Planning obligations*, HMSO.
European Environmental Advisory Councils (1997) *Statement requesting radical reform of the CAP*, The Councils.
Fairlie, S. (1996a) More cheers for pointers to sustainable countryside, *Planning*, 11 October, pp. 26-27.
Fairlie, S. (1996b) *Low Impact Development- Planning and people in a sustainable countryside*, Jon Carpenter.
Fennell, R. (1997) *The Common Agricultural Policy: Continuity and Change*, Clarendon Press.
Ferris-Kaan, R. (ed) (1995) *Managing Forests for Biodiversity*, Forestry Commission.
Fordham, R. (1996) Circular arguments suggest not such affordable housing, *Planning*, 11 October, pp. 24-25.
Forestry Authority (1995) *Creating and managing woodlands around towns*, The Authority.
Forestry Commission (1994) *Annual Report and Accounts for 1993-94*, HC 661 (93-94), HMSO.
Forestry Commission (1995) *Annual Report and Accounts for 1994-95*, HC 749 (94-95), HMSO.
Forestry Commission (1996) *Annual Report and Accounts for 1995-96*, HC 3 (96-97), HMSO.
Franey and Company (1995) *The Planning Factbook*, Gee Publishing.
Freshfields Environment Group (ed) (1996) *Tolley's Environment Law*, Tolley.
Friends of the Earth (1994a) *Working Future? Jobs and the Environment*, The Friends.
---- (1994b) *Losing Interest - A survey of threats to Sites of Special Scientific Interest*, Friends.
Fyson, A. (1995) Route to good practice, *Planning Week*, 17 August, pp. 18-19.
Gardner, B. (1996) *European Agriculture: Policies, Production and Trade*, Routledge.

Gibbard, R. (1995) The Agricultural Tenancies Act is flawed in its every facet. Its a disaster, *Farmers Weekly*, 23 June, p. 91.

Gilg, A. (1996) *Countryside Planning: The first half century*, Routledge.

Gilg, A. and Kelly, M. (1997) Rural Planning in Practice: The case of agricultural dwellings, *Progress in Planning*, 47(2), pp. 75-157.

Gill, P. (1996a) Review of 'Guidelines for landscape and visual impact assessment' by Institute of Environmental Assessment, *Ecos*, 17 (3/4), pp. 95-96.

Gill, P. (1996b) Review of 'Scotland's Mountains: An agenda for sustainable development' by Wightman, *Ecos*, 17 (2), p. 104.

Gill, P. (1996c) Review of Callander (1995) and Worrell and Callander on Scottish forestry (1996), *Ecos*, 17 (2), pp. 103-4.

Grant, M. (1996) *Permitted Development*, Sweet and Maxwell.

Grant, W. (1998) Review of Fennell, *The Common Agricultural Policy*, *Journal of Rural Studies*, 14(3), 1998, pp. 390-1.

Greed, C. (ed) (1996) *Implementing Town Planning The role of town planning in the development process*, Longman.

Green, B. (1997) *Countryside Conservation*, E & FN Spon.

Green, M. (1996) Welsh Rural White Paper, *Ecos*, 17(2), p. 77.

Haigh, N. (ed) (1995) *Manual of Environmental Policy: The EC and Britain*, Cartermill, with 1996 and 1997 updates.

Hams, T., Jacobs, M., Levett, R., Lusser, H., Morphet, J. and Taylor, D. (1994) *Greening your local authority*, Longman.

Harrison, C. (1996) Reconciling leisure uses of the countryside and their environmental impact, *Area*, 28, pp. 339-46.

Harrison, J. (1997) All with the very best of intentions, *Planning*, 13 June, p. 14.

Harvey, G. (1997) *The Killing of the Countryside*, Cape.

Hayton, K. (1996) An advance in Scottish local planning, *Town and Country Planning*, April, pp. 122-124.

---- (1997) Rearrangement of the flawed PAN, *Planning*, 30 May, p. 10.

Healey, P. (1997) Negotiating development, *Town and Country Planning*, April, p. 100.

Henderson, R. (1995) The Environment Act 1995, *Scottish Planning and Environment Law*, 51, pp. 86-87.

Hillman, M. (1995) The Royal Commission on Environmental Pollution's 'Transport and Environment' Report, *Town and Country Planning*, November, pp. 302-303.
Hirst, C. (1997) Hedgerows to enjoy enhanced status, *Planning*, 27 March, p. 1.
Hodge, I. (1996) The Rural White Paper and the Assumptions of Rural Policy, *Journal of Rural Studies*, 12 (4), pp. 331-7.
Hoggart, K., Buller,H. and Black,R. (1995) *Rural Europe: Identity and Change*, Arnold.
Holmans, A. (1995) *Housing demand and need in England 1991-2011*, York Publishing Services.
House Builders Federation (1995) *The Limits of Environmental Capacity*, The Federation.
House of Commons Agriculture Committee (1995a) *Report on MAFF/Intervention Board Departmental Report*, HC 478 (94-95), HMSO. Government reply in HC 776 (94-95).
---- (1995b) *Milk Trading Quota*, HC 512 (94-95), HMSO. Government reply in HC 777 (94-95).
---- (1995c) *Horticulture*, HC 61 (94-95), HMSO. Government reply in HC 778 (94-95).
---- (1996) *The UK Dairy Industry and the CAP Dairy Regime*, HC 40 (95-96), HMSO. Government reply in HC 708 (95-96).
---- (1997) *Environmentally Sensitive Areas and Other Schemes under the Agri-Environment Regulation*, HC 45 (96-97), HMSO.
House of Commons Committee of Public Accounts (1995) *Protecting and managing Sites of Special Scientific Interest*, HC 375(94-95), HMSO.
---- (1996) *Protecting and presenting Scotland's heritage properties*, HC 233 (95-96), HMSO.
House of Commons Defence Committee (1995) *The Defence Estate*, HC 67 (94-95), HMSO.
House of Commons Environment Committee (1994a) *Shopping Centres and their future*, HC 359(93-94), HMSO.
---- (1994b) *Protecting and managing Sites of Special Scientific Interest*, HC 705(93-94), HMSO.
---- (1995a) *This Common Inheritance*, HC 443 (94-95), HMSO.
---- (1995b) *Report on DOE Annual Report for 1995*, HC 518 (94-95), HMSO.

---- (1995c) *The environmental impact of leisure activities*, HC 246 (94-95), HMSO.
---- (1996a) *Rural England White Paper*, HC 163 (95-96), HMSO.
---- (1996b) *Report on DOE Annual Report for 1996*, HC 382 (95-96), HMSO.
---- (1996c) *Water conservation and supply*, HC 437 (95-96) *Interim Report* and HC 42 (96-97) *Final Report*, HMSO.
---- (1996d) *Housing Need*, HC 22 (95-96), HMSO.
---- (1997a) *Countryside Commission: Minutes of Evidence*, HC 41-I (96-97), HMSO.
---- (1997b) *Shopping Centres*, HC 210 (96-97), HMSO.
House of Commons Scottish Affairs Committee (1997) *The future of Scottish agriculture*, HC 629 (95-96), HMSO.
House of Commons Select Committee on European Legislation (1995) *Review of European Legislation*, HC 70 (94-95), HMSO.
---- (1996) *Review of European Legislation*, HC 51 (95-96), HMSO.
---- (1997) *Review of European Legislation*, HC 36 (96-97), HMSO.
House of Commons Trade and Industry Committee (1995) *Regional Policy*, HC 356 (94-95), HMSO.
House of Commons Welsh Affairs Committee (1997) *The preservation of historic buildings and ancient monuments in Wales*, HC 250 (96-97), HMSO.
House of Lords Select Committee on the European Communities (1995) *European Environment Agency*, HL 29 (94-95), HMSO.
---- (1996a) *Enlargement and Common Agricultural Policy Reform*, HL 92 (95-96), HMSO.
---- (1996b) *Reform of the EC Fresh Fruit and Vegetable Regime*, HL 18 (95-96), HMSO.
---- (1997) *Implementation of the Directive on Freedom of Access on the Environment*, HL 9 (96-97), HMSO.
House of Lords Select Committee on Sustainable Development (1995) *Report on the Government's Strategy for Sustainable Development*, HL 72 (94-95), HMSO.
Housing Corporation (1995) *Housing in Rural England*, The Corporation.
Huggett, D. (1996) Still no coastal strategy, *Ecos*, 17 (1), pp. 63-64.

Institute of Environmental Assessment and the Landscape Institute (1995a) *Guidelines for landscape and visual impact assessment*, E & FN Spon.
---- (1995b) *Guidelines for baseline ecological assessment*, E & FN Spon.
Institute of Terrestrial Ecology (1995) *Inland water bodies*, Department of the Environment.
Intervention Board Executive Agency (1994) *Annual Report and Accounts 1993-94*, HC 50 (94-95), HMSO.
---- (1995) *Annual Report and Accounts 1994-95*, HC 666 (94-95), HMSO.
---- (1996) *Annual Report and Accounts 1995-96*, HC 589 (95-96), HMSO.
James, N. (1994) Landmark season on transport, *Ecos*, 15 (3/4), pp.81-82.
Johnston, B. (1997a) In pursuit of a living countryside, *Planning*, 28 February, p.12.
---- (1997b) Advice puts obligations in a more positive light, *Planning*, 14 February, pp. 8,9 and 29.
Joint Nature Conservation Committee (1994) *A UK plant conservation strategy*, The Council.
---- (1997) *Indirect Effects of Pesticides on Birds*, The Committee.
Kirkby, J., O'Keefe, P. and Timberlake, L. (eds) (1995) *The Earthscan Reader in Sustainable Development*, Earthscan.
Labour Party (1995) *A choice for England*, The Party.
Land Use Consultants (1995a) *Effectiveness of Planning Policy Guidance Notes*, Department of the Environment.
---- (1995b) *A study of appeal decisions in Areas of Outstanding Natural Beauty*, Countryside Commission.
---- (1996) *Defining the possible economic effects of Draft PPG13*, Department of the Environment.
Lee, H. (1996) Review of 'Regenerating agriculture' by Pretty, J. in *Ecos*, 17(2), pp. 100-101.
Leeson, J. (1995) *Environmental Law*, Pitman.
Levett, R. (1996) Linking the indicators, *Town and Country Planning*, December, pp. 327-330.
---- (1997) Choose your fantasy, *Town and Country Planning*, April, pp. 98-99.

Lichfield, N. (1996) *Community Impact Evaluation: Principles and Practice*, UCL Press.
Liniado, M. (1996) *Car culture and countryside change*, National Trust.
Lloyd, M. (1996) New Policy Priorities for Scotland, *Scottish Planning and Environment Law*, 54, pp. 29-30.
Local Government Management Board (1995a) *Indicators for Local Agenda 21*, The Board.
---- (1995b) *Sustainable Settlements -A guide for planners, designers and developers*, The Board.
Lowe, P., Marsden, T. and Whatmore, S. (eds) (1994) *Regenerating agriculture*, David Fulton.
Lowe, P., Rutherford, A. and Baldock, D. (1996) Implications of the Cork Declaration, *Ecos*, 17, (3/4), pp. 42-45.
Lucarelli, M. (1995) *Lewis Mumford and the Ecological Region* Guilford Press.
Mackay, D. (1995) *Scotland's Rural Land Use Agencies*, Scottish Cultural Press.
Maddison, D., Pearce, D., Hohansson, O., Calthrop, E., Litman, T. and Verhoef, E. (1996) *Blueprint 5: The True Costs of Transport*, Earthscan.
Mather, A. and Chapman, K. (1995) *Environmental Resources*, Longman.
Mattingley, A. (1994) Aggravated trespass, *Ecos*, 15(3/4), pp. 80-81.
McDonald, R. (1995) *Routes to Rural Opportunities*, Oxford Brookes University.
McDougal, T. (1996a) Call to aim for rural development rather than direct support, *Farmers Weekly*, 15 November, p. 12.
---- (1996b) Stewardship is OK for farm incomes, *Farmers Weekly*, 12 January, p. 10.
---- (1996c) Roaming policy makes the way ahead unclear, *Farmers Weekly*, 24 May, p. 18.
---- (1997a) Countryside Plans of the big three, *Farmers Weekly*, 11 April, pp. 11-13.
---- (1997b) Going organic has never been so popular, *Farmers Weekly*, 4 April, p. 14.
---- (1997c) Boost for wildlife with first UK arable incentive scheme, *Farmers Weekly*, 7 February, p. 12.

McEldowney, J. and McEldowney, S. (1996) *The environment and the law*, Longman.
Milne, R. (1994) Land is the problem for green indicators, *Planning*, 16 December, p. 17.
---- (1996a) Green indicators project opens up public debate, *Planning*, 5 January, p. 10.
---- (1996b) Water wars set to continue as department passes the buck, *Planning*, 25 October, p. 25.
---- (1997a) Sandwell incident highlights problems in information law, *Planning*, 17 January, p. 10.
---- (1997b) Radical rethink in prospect as survey backs plan led, *Planning*, 17 January, p. 6.
Minay, C. (1995) *Reviving Small Communities: A Directory of Projects*, Oxford Brookes University.
Mitchell, C., Cooper, R., Ball, J., Gunneberg, B. and Swift, J. (1995) *The Private Woodlands Survey*, Forestry Commission.
Moffat, I. (1997) Review of Mackay, D., 'Scotland's Rural Land Use Agencies', *Scottish Geographical Magazine*, 113(1), pp. 60-61.
Moir, J. and Watt, A. (1996) Structural problems overlooked in rural patchwork, *Town and Country Planning*, September, pp. 252-254.
Morgan, P. and Nott, S. (1995) *Development control: Law Policy and Practice*, Butterworths.
Morris, P. and Therivel, R. (eds) *Methods of Environmental Impact Assessment*, UCL Press.
Morrison, M. (1995) Great white hope for rural blueprint, *Planning Week*, 25 May, pp. 10-11.
Mullaney, A. and Pinfield, G. (1996) No indication of quality or equity, *Town and Country Planning*, May, pp. 132-133.
Murdoch, J. (1997) The shifting territory of government: some insights from the Rural White Paper, *Area*, 29, pp. 109-118.
Murdoch, J. and Marsden, T. (1994) *Reconstituting Rurality*, UCL Press.
National Farmers Union (1995) *Taking Real Choices Forward*, The Union.
National Heritage, Department of (1996) *Protecting our heritage*, Heritage.
National Trust (1995) *Linking People and Place*, The Trust.

Newson, M., Marvin,S. and Slater, S. (1996) *Pooling our resources*, Council for the Protection of Rural England.
OECD (1994) *The United Kingdom's Environmental Performance*, HMSO.
Office of Population Censuses and Surveys (1995) *Subnational population projections from mid 1993: Series PP3*, HMSO.
Oldfield King Planning (1996) *Rural needs in local authority housing strategies*, Rural Development Commission.
Owen, S. (1995) Review of 'Leisure, Culture and the English Countryside', *Town Planning Review*, 66, pp. 333-334.
Owens, S. and Cowell, R. (1996) *Rocks and hard places*, Council for the Protection of Rural England.
Pennington, M. (1996) *Conservation and the countryside: By Quango or Market?*, Institute of Economic Affairs.
Perraton, J. (1996) Policy note floats a vision of a very pale green countryside, *Planning*, 4 October, p. 29.
Pinfield, G. (1994) 'Must have' guides and a moving target, *Town and Country Planning*, October, pp. 287-288.
Pitt, J. (1995) Harmonious note for green belt forestry?, *Planning*, 3 March, p. 10.
Planning Inspectorate Executive Agency (1995a) *Annual Report for 1994-95*, HMSO.
---- (1995b) *Business and Corporate Plan 1995-99*, HMSO.
---- (1996) *Annual Report for 1995-96*, HMSO.
Potter, C. (1995) Tomorrow's Countryside, *Ecos*, 16 (3/4), pp. 7-9.
---- (1996) Review of Environment Committee's report on Rural England: The Rural White Paper, *Ecos*, 17 (2), p. 93.
---- (1997) Review of Harvey, G., 'The Killing of the Countryside' *Ecos*, 18(2), pp. 102-103.
Pretty, J. (1995) *Regenerating agriculture: Policies and practice for sustainability and self-reliance*, Earthscan.
Pritchard, S. (1995) Too many paws in the rural honeypot, *Planning Week*, 27 July, pp. 10-11.
Reid, C. (1995) Cairngorms partnership, *Scottish Planning and Environmental Law*, 47, pp. 2-3.
Rodgers, C. (ed) (1996) *Nature Conservation and Countryside Law*, University of Wales Press.

Roger Tym and Partners with Pagoda Associates (1995) *Review of the implementation of PPG16*, English Heritage.

Ross-Robertson, A. (1997) Natural Priorities for Scottish Natural Heritage, *Scottish Planning and Environment Law*, 59, pp. 2-3.

Royal Commission on Environmental Pollution (1997) *Sustainable Use of Soil*, Cm 3165, HMSO.

Royal Society for the Protection of Birds (1994) *Environmental measures*, The Society.

---- (1995) *Biodiversity Challenge: An Agenda for Conservation Action in the UK*, First and Second editions, The Society.

---- (1996a) *A Step by Step Guide to Environmental Appraisal*, The Society.

---- (1996b) *Wildlife Impact-The Treatment of Nature Conservation in Environmental Assessment*, Report by David Tyldesley & Associates, The Society.

---- (1996c) *A review of Indicative Forestry Strategies*, The Society.

---- (1996d) *Natural conditions*, The Society.

---- (1996e) *High and Dry*, The Society.

Royal Town Planning Institute, Royal Institution of Chartered Surveyors, the County Planning Officers Society and the District Planning Officers Society (1995) *Tomorrow's countryside: A rural strategy*, The Institute.

Rudlin, D. and Falk, N. (1995) *21st Century Homes: Buildings to Last*, Urbed.

Rural Development Commission (1995a) *Section 106 Agreements and Private Finance for Rural Housing Schemes*, The Commission.

---- (1995b) *Tourism in the Countryside*, The Commission.

---- (1995c) *Rural Economic Activity*, The Commission.

---- (1995d) *Planning for People and Prosperity*, The Commission.

---- (1996a) *The employment impact of changing agricultural policy*, The Commission.

---- (1996b) *Fair shares for rural areas? An assessment of public resource allocation systems*, The Commission.

---- (1996c) *Rural Transport-The vital link*, The Commission.

---- (1996d) *Community and Voluntary Transport in Rural England*, The Commission.

---- (1996e) *Old Buildings New Opportunities*, The Commission.

---- (1997a) *Telecommunications Development in Rural England*, The Commission.
---- (1997b) *Teleworking and Rural Development*, The Commission.
Rusling, R. (1997) New regs for hedges, *Farmers Weekly*, 13 June, p. 94.
Scottish Natural Heritage (1994) *Agriculture and the Natural Heritage*, Heritage.
---- (1995a) *The Natural Heritage of Scotland-an Overview*, Heritage.
---- (1995b) *The Environment-Who Cares*, Heritage.
---- (1996) *Natural Priorities*, Heritage.
---- (1997) *Long Distance Routes in Scotland: An SNH policy statement*, Heritage.
Scottish Office (1996a) *Evaluation of the land cover of Scotland 1998 project*, HMSO.
---- (1996b) *Guide to measures available to control the recreational use of water*, HMSO.
---- (1996c) *Living in Rural Scotland*, HMSO.
---- (1996d) *Scoping study on rural development in Scotland*, HMSO.
---- (1997a) *Evaluation of Indicative Forestry Strategies*, The Office.
---- (1997b) *New Ideas in Rural Development 3: Involving rural communities: the CADISPA approach*, HMSO.
Scottish Wildlife and Countryside Link (1994) *Shaping the new councils: Shaping the countryside*, Link.
Secretary of State for Wales (1995) *The Environmental Agenda for Wales*, Welsh Office.
Secrett, C. (1996) *Defining sustainable development in the transport sector*, UK Round Table on Sustainable Development.
Segal Quince Wicksteed (1996) *The Impact of tourism on rural settlements*, Rural Development Commission.
Selman, P. (1996) *Local Sustainability: Managing and Planning Ecologically Sound Places*, Paul Chapman.
Shucksmith, M., Chapman, P., Clark, G., Black, S. and Conway, E. (1996) *Rural Scotland Today-The best of both worlds*, Avebury.
Shucksmith, M., Roberts, D., Scott, D., Chapman, P. and Conway, E. (1997) *Disadvantage in Rural Areas*, Rural Development Commission.

Sidaway, R. (1996) Linking which people to whose place?, *Ecos*, 17 (1), pp. 56-58.

Speer, R. and Dade, M. (1995) *How to stop and influence planning permission*, JM Dent.

Spray, M. (1994) Welsh Agency cut back for carve-up, *Ecos*, 15 (3/4), pp. 73-74.

Standing Advisory Committee on Trunk Road Assessment (1994) *Trunk Roads and the Generation of Traffic*, HMSO.

Steel, J. (1995) *Public Access to Information: An evaluation of the Local Government (Access to Information) Act 1985*, BEBC.

Steeley, G. (1995) A guide through the maze, *Town and Country Planning*, September, pp. 240-241.

Swales, V. (1995) Environmental land management schemes, *Ecos*, 16 (2), pp. 69-70.

Tayhoe, D. (1997) Countryside movement dissolves, *Ecos*, 18 (1), pp. 90-91.

Taylor, A., Gordon, J. and Usher, M. (eds) (1996) *Soils, Sustainability and the Natural Heritage*, HMSO.

Taylor, D. (1994) PPG9: A lawyer's dream?, *Planning Week*, 3 November, p. 9.

Taylor, D. (1996) New developments in Green Belt guidance, *Planning Week*, 2 February, p. 9.

Technology Foresight Panel, Office of Science and Technology, Cabinet Office (1995) *Agriculture, natural resources and the environment*, HMSO.

Tewdwr-Jones, M. (ed) (1996) *British planning policy in transition: Planning in the 1990s*, UCL Press.

Thompson, S. (1997) Fight is on for strategic assessment, *Ecos*, 18(1), pp. 87-88.

Tomalin, C. (1996) Revised Planning Policy Guidance for Town Centres and Retail Development, *Town Planning Review*, 67, pp. iii-vi.

Tompsett, R. (1995) Environmentalists set targets for the biodiversity challenge, *Planning Week*, 26 January, p. 11.

Toogood, M. (1995) Biodiversity Challenge - wither the wider agenda? *Ecos*, 16(1), pp. 65-66.

Town and Country Planning Association (1996) *The People-Where will they go?*, The Association.

Traill-Thomson, J. (1996) *Rural Futures: Issues for the forthcoming White Paper*, Exeter University.
Transport, Department of (1996) *The valuation of environmental externalities*, HMSO.
Transport Studies Group (1996) *Public attitudes to transport policy and the environment*, University of Westminster.
Trench, S. and Oc, T. (1995) *Current Issues in Planning*, Avebury.
Tyldesley, D. (1994) PPG9 Nature Conservation, *Ecos*, 15 (3/4), pp. 79-80.
---- (1996) Assessment process faces wildlife impact, *Planning*, 2 February, pp. 8-9.
UK Round Table on Sustainable Development (1997) *Getting Around Town*: A report by J. Adams, available from the Department of the Environment.
United Kingdom Climate Change Impacts Review Group (1996) *Review of the potential effects of climate change to the UK*, HMSO.
Ward, C. (1997) Hedging bets over hedges, *Planning*, 16 May, p. 15.
Watkins, C. (ed) (1996) *Rights of Way-Policy, Culture and Management*, Pinter.
Welsh Office (1996) *Planning Guidance (Wales) Planning Policy*, HMSO.
Wightman, A. (1996) *Scotland's Mountains: An Agenda for Sustainable Development*, Scottish Wildlife and Countryside Link.
Williams, N., Shucksmith, M., Edmond, H. and Gemmell, A. (1996) *Scottish Rural Life Update: A revised socio-economic profile of Rural Scotland*, HMSO.
Winter, M. (1996) *Rural Politics: Policies for Agriculture, Forestry and the Environment*, Routledge.
Winter, M., Gasson, R., Curry, N., Selman, P. and Short, C. (1996) *Socio-economic evaluation of free conservation advice provided to farmers in England by ADAS and FWAG*, Countryside and Community Press.
Winter, M., Watkins, C. and Cox, G. (1996) *Game management in England: Report of a socio-economic survey*, Countryside and Community Press.
World Wide Fund for Nature (1994) *Green Gauge*, The Fund.

World Wide Fund for Nature UK (1996) *Green Gauge '96*, New Economics Foundation, CPRE, FoE, RSPB, The Wildlife Trusts and World Wide Fund for Nature UK.

Worrell, R. and Callander, R. (1996) *Native woodlands and forestry policy in Scotland: A discussion paper*, Native Woodlands Policy Forum.

WS Atkins and MCL Transport (1996) *Community and Voluntary Transport in Rural Areas*, Rural Development Commission.

Young, J. (1995) Landowners appeal for ministry of rural affairs, *The Times*, 15 February.

Zetter, R. (1997) *The role of elected members in Plan Making and Development Control*, Royal Town Planning Institute.

For Product Safety Concerns and Information please contact our EU representative GPSR@taylorandfrancis.com Taylor & Francis Verlag GmbH, Kaufingerstraße 24, 80331 München, Germany

Printed and bound by CPI Group (UK) Ltd, Croydon, CR0 4YY
14/09/2026
02216223-0011